T0291530

Africa's Gene Revolution

Africa's Gene Revolution

Genetically Modified Crops and the Future of African Agriculture

MATTHEW A. SCHNURR

McGill-Queen's University Press
Montreal & Kingston • London • Chicago

ISBN 978-0-7735-5903-5 (cloth)
ISBN 978-0-7735-5904-2 (paper)
ISBN 978-0-2280-0044-0 (ePDF)
ISBN 978-0-2280-0045-7 (ePUB)

Legal deposit fourth quarter 2019
Bibliothèque nationale du Québec

Printed in Canada on acid-free paper that is 100% ancient forest free
(100% post-consumer recycled), processed chlorine free

This book has been published with the help of a grant from the Canadian
Federation for the Humanities and Social Sciences, through the Awards
to Scholarly Publications Program, using funds provided by the Social Sciences
and Humanities Research Council of Canada.

We acknowledge the support of the Canada Council for the Arts.

Nous remercions le Conseil des arts du Canada de son soutien.

Library and Archives Canada Cataloguing in Publication

Title: Africa's gene revolution: genetically modified crops and the future
    of African agriculture / Matthew A. Schnurr.

Names: Schnurr, Matthew A., author.

Description: Includes bibliographical references and index.

Identifiers: Canadiana (print) 20190137290 | Canadiana (ebook) 20190137339 |
    ISBN 9780773559042 (paper) | ISBN 9780773559035 (cloth) |
    ISBN 9780228000440 (ePDF) | ISBN 9780228000457 (ePUB)

Subjects: LCSH: Transgenic plants—Africa. | LCSH: Crops—Genetic
    engineering—Africa. | LCSH: Agriculture—Africa.

Classification: LCC SB123.57 .S36 2019 | DDC 631.5/233096—dc23

This book was typeset by Marquis Interscript in 10.5 / 13 Sabon.

Contents

Acknowledgments

This book is the culmination of fourteen years of research into agricultural technology in sub-Saharan Africa. There are many people to thank.

Undertaking farm-level research in Africa is expensive work. Grants from the John Templeton Foundation and the Social Sciences and Humanities Research Council of Canada enabled me to travel frequently and spend extended periods of time on the continent. The final part of the manuscript was completed as part of a six-month fellowship at the Stellenbosch Institute for Advanced Study (STIAS) in South Africa. The insights that I gleaned during my time at STIAS greatly enhanced the value of the manuscript as a whole.

I was very lucky to land an academic position amongst a group of supportive colleagues in the Department of International Development Studies at Dalhousie University. John Cameron, Bob Huish, Nissim Mannathukkaren, Ajay Parasram, and Theresa Ulicki are all terrific scholars and, more importantly, generous and genuine people. I have benefited enormously from exchanges and collaborations with other social scientists who share similar interests in African agricultural development, especially Lincoln Addison, Rachel Bezner Kerr, Brian Dowd-Uribe, Dominic Glover, Chris Gore, Ann Kingiri, Bill Moseley, Sarah Mujabi-Mujuzi, Rachel Schurman, Sarah Ssali, and Glenn Stone. Special thanks to Lincoln, Brian, and Sarah M., who agreed to let me reproduce parts of our collaborative work in these pages. Fazeela Jiwa provided generous and insightful comments on a preliminary draft that greatly enhanced the final product. Correy Baldwin and Shannon Fraser were meticulous in their copy-editing, proofreading, and indexing. I am very grateful to Jacqueline Mason at McGill-Queen's, who first broached the idea of undertaking a book and stuck with me during the years it took to see it through.

The fieldwork upon which this book is based was supported by an exemplary group of research assistants who added value to each and every page. Sarah Mujabi-Mujuzi and I have been collaborating for ten years and I hope we will continue doing so for the rest of my days. Rodgers Atwooki manages much of the logistics associated with research in Uganda. He is a true friend and knows the back roads of Kampala better than anyone. Emmanuel Nsereko's guest house in Kampala served as my home away from home. Tonny Miiro in Uganda and Hamadou Dioff in Burkina Faso were generous with their expertise and their time. I am extremely grateful to the hundreds of farmers who took the time to share their knowledge and perspectives with me. I hope that you will hear your voices reflected in these pages.

In Canada I have benefited greatly from a stream of talented research assistants who have helped me track down all sorts of difficult-to-find information. Thanks to Lucy Hinton, Rachel Matheson, Kelly Pickerill, Annie Hinton, Alysia Strobl, and Meredith Nelson for all their work on my behalf. Meghan Ruston is responsible for the excellent figures that appear throughout. Alanna Taylor has served as my research coordinator for the past four years and has played an important role in the manuscript's evolution. Her dedication, resourcefulness, and attention to detail have made the final product infinitely better than it would have been otherwise.

My family has been a constant source of support and encouragement. My parents Brian and Annalee, my sister Jessica and bonus brother Mike (who designed the book cover!), and my family-in-law Margo, Frank, and Lisa have served as unfailing backers of this project from beginning to end. My grandparents, Irving and Florence Rother, were lifelong boosters. They got to see much of the process that went into this book, but sadly did not live to see the final product. I know they would be proud of both. My wife, Natalie, and my sons, Louis and Evan, believed in me even when I doubted myself. I dedicate this book to them.

Abbreviations

AATF	African Agricultural Technology Foundation
ABNE	African Biosafety Network of Expertise
ABSP	Agricultural Biotechnology Support Project
AGERI	Agricultural Genetic Engineering Research Institute
AGRA	Alliance for a Green Revolution in Africa
ASARECA	Association for Strengthening Agricultural Research in Eastern and Central Africa
AU	African Union
BBW	banana bacterial wilt
BMGF	Bill and Melinda Gates Foundation
Bt	*Bacillus thuringiensis*
CAADP	Comprehensive Africa Agriculture Development Program
CABIO	Collaborative Agricultural Biotechnology Initiative
CDO	Cotton Development Organization
CFT	confined field trial
CGIAR	Consultative Group on International Agricultural Research
CIMMYT	International Maize and Wheat Improvement Center (Centro Internacional de Mejoramiento de Maíz y Trigo)
COMESA	Common Market for Eastern and Southern Africa
CRISPR	Clustered Regularly Interspaced Short Palindromic Repeats
CSO	civil society organization
DTMA	Drought Tolerant Maize for Africa
FAO	Food and Agriculture Organization

GM	genetically modified
GMO	genetically modified organism
HT	herbicide tolerant
IFPRI	International Food Policy Research Institute
INERA	Institut de l'environnement et recherches agricoles
IRMA	Insect Resistant Maize for Africa
ISAAA	International Services for the Acquisition of Agri-biotech Applications
KARI	Kenya Agricultural Research Institute
KALRO	Kenya Agricultural and Livestock Research Organization (formerly KARI)
NARO	National Agricultural Research Organisation (Uganda)
NARS	National Agricultural Research Systems
NBA	National Biosafety Authority
NBC	National Biosafety Committee
NEPAD	New Partnership for Africa's Development
OPVS	open-pollinated varieties
P3S	public-private partnerships
PBS	Program for Biosafety Systems
SADLF	Southern African Drought and Low Soil Fertility Project
SAGENE	South African Committee for Genetic Experimentation
SFSA	Syngenta Foundation for Sustainable Agriculture
SOFITEX	Société burkinabè des fibres textiles
STMA	Stress Tolerant Maize for Africa
UNEP-GEF	United Nations Environment Programme's Global Environment Facility
USAID	United States Agency for International Development
WEMA	Water Efficient Maize for Africa

Africa's Gene Revolution

A Political Ecology
of Africa's Gene Revolution

African agriculture is much maligned. For hundreds of years colonial and then development experts have portrayed it as backwards, inefficient, and unproductive. Statistics reinforce this depiction. Production levels for African smallholder farmers have remained stagnant for decades, with actual yields of staple crops achieving less than a quarter of their potential.[1] According to the United Nations' Food and Agriculture Organization (FAO), over the latter part of the twentieth century, yields for cereal crops increased by 67 per cent. During this same period, cereal yields in Asia increased by 156.5 per cent and in South America by 120.5 per cent. Per capita food production across Africa remains nearly the same today as it was in 1960, while over this same time period increases of 102 per cent were recorded in Asia and 63 per cent in Latin America.[2] And while agricultural productivity in sub-Saharan Africa (SSA) has gradually increased since the mid-1980s, the productivity gap between SSA and other parts of the developing world continues to widen, with SSA achieving productivity growth that is only 61 per cent of what is achieved elsewhere.[3] SSA's average cereal yields, for example, are approximately half of those achieved in Asia and Latin America.[4]

The most powerful rhetorical device for representing the underutilized potential of African agriculture is that of the yield gap, which refers to the differential between actual yields obtained by farmers and those possible under ideal conditions. These data suggest that Africa continues to lag behind relative to agricultural production in the rest of the world. Average grain yields hover around one-half to one-third of the global average.[5] The gap between potential and realized yields for rice is the largest in the world.[6]

In his 2015 annual letter to his and his wife's philanthropic foundation, Bill Gates explains passionately the burden of Africa's persistent yield gaps. Gates's perspective looms large in these debates, since his foundation now represents the single largest donor in international development, and plays a foundational role in shaping and implementing agricultural development programs across Africa. Gates is especially focused on the yield gap for maize, the most important cereal crop on the continent. He laments the fact that maize yields in Africa are only a fraction of what they are in more sophisticated farming systems: maize yields in the United States are more than five times that obtained by the average African farmer. Gates then proceeds to lay out his big bet for fixing these discrepancies: "The world has already developed better fertilizer and crops that are more productive, nutritious, and drought- and disease-resistant; with access to these and other existing technologies, African farmers could theoretically double their yields." The Bill and Melinda Gates Foundation (BMGF) is aiming to increase productivity by more than 50 per cent in fifteen years. According to Gates, this investment in increasing productivity "will make a world of difference for millions of farmers."[7]

Gates's big bet hinges largely on the adoption of productivity enhancing technologies such as enhanced seeds to redress this persistent yield gap. Genetic modification is a breeding technology that allows molecular scientists to alter a plant's genetic code in order to amplify a beneficial trait or minimize a harmful one.[8] It is hard to find someone who does not have strong opinions on genetic modification. Many biotech boosters hold up genetic modification as a promising strategy for sustainably intensifying production rates and overcoming the persistent yield gap that plagues Africa.[9] Development organizations such as the FAO and World Bank argue that higher yields via genetically modified (GM) crops will increase incomes, mitigate hunger, and make farming practices more sustainable across Africa.[10] Gates himself reflected on the potential of genetic modification at the World Economic Forum in 2016, saying, "it's pretty incredible because it reduces the amount of pesticide you need, raises productivity [and] can help with malnutrition by getting vitamin fortification … if you want farmers in Africa to improve nutrition and be competitive on the world market, you know, as long as the right safety things are done, that's really beneficial. It's kind of a second round of the green revolution. And so the Africans I think will choose to let their people have enough to eat."[11] Of course, genetic modification has just as many opponents, who decry it as a tool of industrial agriculture, one that increases corporate control over seeds and deepens farmer dependence on fertilizers and pesticides. Others argue that the advance of genetic modification in Africa is part of a nefarious plot on the part of multinational corporations to foster global demand for their products.[12]

Much is at stake in debates over yields and technology. Agriculture is more important in Africa than in almost any other part of the world, where it accounts for between 25 and 40 per cent of the continent's gross domestic product (GDP) and employs over 65 per cent of its labour force.[13] Nearly 50 per cent of the continent's 1.2 billion people rely on agriculture to meet their daily needs.[14] Growth regenerated via the agricultural sector is thus viewed as one of the most effective mechanisms for reducing poverty and hunger in Africa.[15]

Agriculture's role as the continent's growth engine seems set to continue. SSA is home to the fastest growing population in the world, projected to reach 1.5 billion people by 2050.[16] To keep pace, the World Bank estimates that food production in the region will need to increase by more than 60 per cent from current levels. Investment in smallholder production will be critical, as these producers account for 80 per cent of farms in the region.[17] Climate change presents an ever-present danger: the Intergovernmental Panel on Climate Change identified Africa as the most vulnerable to the effects of climate change, largely due to its low adaptive capacity, predicting that by 2020 yields from rain-fed agriculture could be reduced by up to 50 per cent. Genetic modification is being advanced as a means of redressing stagnating yields, of alleviating poverty and hunger, of staving off the threat of population growth, and of mitigating the unpredictable effects of climate change.

This book examines the potential of genetic modification to achieve these goals. It surveys the gamut of GM crops that have already been released alongside a range of experimental varieties that are currently being developed, looking backwards and forwards in order to offer a systematic assessment of the technology's capacity to enhance agricultural production across the continent. I pay particular attention to the fit between technology and context; my aim is to eschew the more polarized and politicized elements of this debate by evaluating the impact of each specific GM variety within the agricultural system it is designed to enhance. The book's main argument is rooted in this question of fit: my analysis reveals that Africa's Gene Revolution does not sufficiently take into consideration smallholder farmers' lived realities in the form of their geographic, social, ecological, and economic contexts. My analysis reveals a profound incongruence between the optimistic rhetoric that surrounds GM technology and the realities of the smallholder farmers who are its intended beneficiaries.

SMALLHOLDER FARMERS IN AFRICA

Peter Musisi is a forty-five-year-old farmer in Semuto Subcounty, Nakaseke District, located just over sixty kilometres northwest of Kampala. His compound is clean and inviting. He has a well-kept lawn shaded by a twenty-year-old ficus tree, with large clearings that allow for the drying of crops after they are harvested. He lives with his wife, Ephrance, and twelve additional family members: seven of his own children, two from a brother-in-law who passed away, and three from a sister who resides in Kampala.

Peter and Ephrance farm a combined four hectares of land on which they grow a mix of the region's staple crops, including cooking banana, maize, beans, groundnuts, squash, sweet potato, and various vegetables. They farm for their own consumption, selling off the surplus to middlemen who use old motorcycles to ferry their produce towards larger markets in Kampala. Peter's operations are decidedly small-scale. He recycles and replants most of the seed he uses, exchanging some with neighbours or, rarely, purchasing new planting materials via private nurseries. He uses cow manure and limited amounts of synthetic fertilizers, but lacks the income to invest in large amounts of herbicides or pesticides. When labour demands are high, he enlists his children to help, as he generally lacks funds to hire outside workers. He is eager to receive information and training on new tools and techniques from agricultural extension workers, but these resources seldom reach his farm. Instead, most of his farming information comes from neighbours and local radio programming. To overcome these limitations in inputs and information, Peter spearheaded the formation of a local farmers association in 2005, which now serves primarily to pool resources and leverage better prices when selling surplus crops.

Peter's farming system is typical of smallholder farmers in Africa, who together produce more than 70 per cent of the continent's total food production.[18] Peter is slightly better off than many of this contemporaries: according to the FAO, over 80 per cent of the farms across the continent have holdings of less than two hectares, while only 3 per cent have holdings above ten hectares.[19] I use the term "smallholder" to designate the farming system that is most prevalent across the continent: land size is limited (typically under three hectares), managed by the household, with limited amounts of external inputs such as synthetic fertilizers, pesticides, or mechanization. Most farms produce for subsistence purposes first, and then sell off any surplus on the local market. Many farms also diversify in terms of their practices, which might include forestry, fisheries, pastoral, and aquaculture production, or cultivate one or more cash crops such as coffee, tobacco, cocoa, or cotton that are used to supplement their livelihoods. Together, smallholder farmers constitute one of the most marginalized and vulnerable categories in the world, comprising half of the world's chronically malnourished people and the majority of those living in extreme poverty.[20]

In many respects, then, Peter and Ephrance are representative of the continent's smallholder farmers, in that they farm on a relatively small piece of land, they have little access to inputs or technologies, and they rely almost exclusively on family labour.[21] But there are some thirty-three million smallholder farmers in Africa. Each household faces it own unique set of climatic, economic, and political circumstances. The description of Peter and Ephrance's particular farming context provided here tells us little about the far-reaching set of circumstances facing their contemporaries across Africa. Peter and Ephrance are unique: their language, their agro-ecological system, their cultural heritage – all of these are distinct from farmers in other parts of the country, let alone the continent. African farmers are a diverse group who defy any singular characterization; as such, this book will zoom in to examine specific African agricultural systems and use these case studies as jumping-off points for the analysis of broader trends, while remaining careful not to essentialize Africa as a homogeneous, undifferentiated whole. At its core, this book is about Peter and Ephrance, the other thirty-three million other African smallholder farmers who work tirelessly to meet their families' needs, and the potential for GM crops to make their lives better.

A (SHORT) HISTORY OF AGRICULTURAL DEVELOPMENT IN AFRICA

An analysis of the potential for a new technological innovation like genetic modification to improve yields and livelihoods for African farmers requires a longer-term understanding of the role agriculture has played in the continent's history. Pre-colonial African agriculture was dominated by subsistence farming amidst local markets, credit schemes, and regular surpluses.[22] Land was abundant, but in most regions the population was sparse and spread out, leading to labour shortages. As a result, land-extensive techniques – techniques that used less labour and more land, like land rotation and shifting cultivation – remained the most common forms of cultivation.[23]

Colonialism completely transformed the way African agricultural systems functioned. What was a fairly diversified, subsistence-oriented production system was overhauled to prioritize commodity crops for export to satisfy the demands of European consumers. Vast tracks of land supporting thousands of homesteads were expropriated, razed, and transformed into monoculture plantations growing highly prized crops such as cocoa, tobacco, sugar, cotton, tea, groundnuts, and coffee. Africans were forced to labour on these farms, subjected to unspeakable violence in order to quell any objection or resistance. In this way African agricultural production was remade to prioritize the outsourcing of food production, through which Europe subsidized imported foods based on the subjugation of African land and labour.[24]

Up until the turn of the millennium most agricultural development initiatives in Africa were tied to World Bank-led structural adjustment programs and poverty reduction strategy papers, which sought to minimize the continent's heavy reliance on agriculture by investing in the production and export of manufactured goods.[25] What little investment in agriculture was maintained during this period was directed primarily towards the expansion of commodity crops destined for the international market, creating a system of production dominated by large-scale plantations that produced single commodities such as coffee, cotton, tobacco, tea, cocoa, sesame, and cashew nuts for export in their raw form, with little investment in value-added processing.

This half century of externally driven agricultural development that had been imposed onto the continent by outsiders was finally rebuffed around the turn of the new millennium, at which point there was a surge of calls for more participatory and country-owned agricultural development strategies. This culminated in the New Partnership for Africa's Development (NEPAD) unveiling its Comprehensive Africa Agriculture Development Program (CAADP) in 2003, which recentred agriculture as the main vessel for African economic development. This African-led initiative pledged ambitious targets for agricultural investment (10 per cent of national budgets) and annual rates of agricultural growth (6 per cent). The stated aim of CAADP was to "stimulate and facilitate increased agricultural performance through improvements in policy and institutional environments, access to improved technologies and information, and increased investment financing," which it aspired to achieve via four key tenets of African ownership, harmonization of policies, evidence-based decision-making, and building partnerships.[26] In order to achieve these lofty goals, CAADP focused heavily on developing strategic partnerships, especially with the private sector,[27] as well as substantial investment in agricultural research and technology dissemination as a key strategy for enhancing productivity.[28] Most of the funding for CAADP programming came from bilateral development agencies such as the United States Agency for International Development (USAID), CIDA, and the United Kingdom's Department for International Development (DFID).

The World Bank's 2008 *World Development Report* cemented CAADP's emphasis on investment, technology, and partnerships as the key to modernizing African agriculture. The starting point for this seminal report is that boosting rates of agricultural productivity is the best way to stimulate growth and redress food insecurity in Africa. Maligning the "still-unsatisfactory" performance of African agriculture, the report calls for a "smallholder-based productivity revolution in agriculture" that would encompass greater investment in agricultural research, extensive services, irrigation, inputs, and credit.[29]

This prominent call for greater agricultural intensification precipitated two of the most significant investments in African agriculture in recent years: Grow Africa and the New Alliance for Food Security and Nutrition. Grow Africa was implemented as a joint initiative of the AU (African Union), NEPAD, and the World Economic Forum in 2011, designed as a "convener and catalyst" of private sector investment in African agriculture.[30] The New Alliance for Food Security was launched in 2012 by the G8 and USAID as a means of building and integrating farmers into value chains, in order to encourage investment and accelerate agriculture-led growth in Africa. Operating in ten African countries with over 200 private sector partners, the New Alliance aims to invest over US$10 billion in African agriculture.[31]

CAADP's emphasis on "strategic partnerships," Grow Africa's "partnership platform," and the New Alliance for Food Security and Nutrition's commitment to "responsible investment" embody the emerging commitment to incorporating the private sector as a key ally in the agricultural development sector. Over the past twenty years, this model of leveraging private sector investment to enhance agricultural yields has emerged as the central platform for improving agricultural development in Africa. This emphasis on agricultural modernization incorporates a strategic focus on connecting producers to markets via value chains as a means of transforming subsistence-oriented farming into commercial, market-oriented production and surplus generation. Ultimately, the goal is to create a more productive, diversified agricultural sector that will increase food availability across the continent, decrease food prices for consumers, and provide new opportunities for agricultural investment via export and agro-processing. This will require fewer people to grow food, enabling thousands to leave agriculture and find employment in other sectors of the economy, primarily in urban centres. According to its proponents, this broad economic transformation offers the best opportunity to alleviate hunger and poverty in Africa.[32]

AFRICA'S GREEN REVOLUTION

The single most transformative event in the realm of agricultural development over the past seventy years was the Green Revolution. The major investment in high-yielding varieties of carbohydrate staple crops, especially wheat, maize, and rice, was buoyed by the idea that increasing agricultural productivity was the most promising means of boosting poor farmers out of poverty. This period of technological investment and transfer was focused on developing high-yielding varieties that were highly responsive to fertilizer, pesticides, and intensive irrigation, with a geographical emphasis on Latin America and South Asia.[33] The Green Revolution was funded primarily by private philanthropists, especially the Rockefeller and Ford Foundations, and facilitated by multilateral development agencies such as the UN's FAO and research centres, catalyzing the emergence of the UN's Consultative Group on International Agricultural Research (CGIAR) network.

The enduring impacts of the Green Revolution remain contentious. Yields for key carbohydrate crops such as wheat and rice nearly doubled in Mexico, India, and Pakistan between 1965 and 1970.[34] According to the FAO, between 1960 and 2000, global yields for wheat, rice, and maize increased by 208 per cent, 109 per cent, and

157 per cent, respectively, due primarily to the enhanced varieties, inputs, and practices that comprised the Green Revolution.[35] The Green Revolution also precipitated a broader investment in rural agricultural development programming, including input subsidies, access to credit, expanding market access, and investments in agricultural extension, which helped to keep food prices low, increase rural incomes, and reduce rural poverty.[36] As a result, many target countries such as Bangladesh, India, and the Philippines underwent a doubling of per capita food production between 1960 and 2000, facilitating their transition from being food insufficient to food exporters.[37] According to proponents, the Green Revolution showcased how improved technologies could be used to transform traditional agriculture into more modern practices, which in turn hastened meaningful and sustained economic growth.

But critics argue that these increases in aggregate production mask the Green Revolution's hazardous legacies, which include intensifying inequities amongst growers by creating a system of high-input agriculture whose benefits were disproportionately captured by larger, wealthier farmers, by entrenching pesticide- and fertilizer-intensive practices that are not sustainable in the long-term, and by consolidating monoculture production among fewer cultivated varieties, which reduces genetic variability, making the crops more vulnerable to pests and diseases.[38] Other longer-term impacts include the expropriation and consolidation of land, as larger land owners sought to gain more control as land prices increased, pushing smaller farmers off the land and transforming them into landless labourers, and the disproportionate capture of benefits by male farmers relative to their female counterparts.[39] The excessive use of agrochemicals needed to sustain these higher-yielding varieties led to environmental deterioration, including decreased soil fertility and water pollution.[40]

While the debate over the implications of the original Green Revolution persists, there is no disputing that Africa was largely bypassed by this period of agricultural innovation and investment. Indeed, Africa's contemporary agricultural problems – particularly the persistence of the yield gap – are often traced back to its having missed out on the agricultural innovations and technological revolutions that propelled the Green Revolution in Asia and Latin America.

This omission was the impetus for the single largest investment in African agriculture in this millennium, known as the New Green Revolution for Africa (GR4A). In 2006, the BMGF and the Rockefeller

Foundation formed the Alliance for a Green Revolution in Africa (AGRA), designed to spur a "uniquely African" Green Revolution. Its mandate is to lead an agricultural transformation in Africa by increasing productivity and profitability of smallholder farmer households.[41] AGRA's work revolves around four values: African led, farmer centred, partnership driven, and integrated.[42] AGRA currently works in eleven African countries alongside dozens of partners including various CGIAR centres, national governments, international organizations, research institutions, national agricultural organizations, and NGOs.[43]

AGRA's principal mission is to redress the entrenched African yield gap: "that is the core of what we do ... we want to close that gap."[44] Its operations are premised on scaling up African agriculture via greater integration of input and output markets.[45] Their chosen mechanism for accomplishing this is to leverage the power of the private sector. Specifically, AGRA is focused on expanding the African-owned private seed sector as the best means of linking farmers with improved seed and inputs. In its first decade of operation, AGRA invested over US$430 million targeting agricultural transformation for thirty million smallholder farmers across eleven target countries.[46] Through this investment, modernization will be the catalyst to "unleash [Africa's] agricultural potential."[47] In this way, the logic of yield enhancement as a stimulus for rural development, which underpinned the original Green Revolution, similarly drives this renewed commitment to sustainable intensification in Africa.

FROM GREEN REVOLUTION TO GENE REVOLUTION

Seeds and technology are at the centre of this vision for a new Green Revolution for Africa; the first guiding principle of AGRA's core programming model is "No improved seed, no Green Revolution."[48] AGRA considers the use of outdated, unproductive technologies to be the biggest barrier in African agriculture. Agricultural modernization will enable farmers to adopt useful and affordable technologies, which will in turn increase yields and protect crops from pests, droughts, and other stresses.[49] Proponents point out that yield increases in the past thirty years have been primarily due to technical advances, based on the Green Revolution's model of sustainable intensification. For example, the doubling of global cereal production between 1970 and 2008 was due primarily to increases in productivity from technological improvements, since the total land under cultivation only increased by 10 per cent.[50]

As the push for a New Green Revolution for Africa began to expand, so too did the enthusiasm for integrating new genome-enhancing technologies as a tool to enhance agricultural productivity. This interest in using genetic modification to attain agricultural development priorities is now a dominant feature of GR4A thinking, which has precipitated the emergence of Africa's Gene Revolution.[51] There are two distinct strands to this approach. First, it is premised upon embedding desired phenotypic changes directly into the seed in order to produce technologies that are scale neutral in the hopes of producing lasting benefits for poor, small-scale farmers. Second, and distinct from the original Green Revolution, the Gene Revolution is underpinned by the involvement of the private sector, which owns and controls most of the genetic constructs that are being imported, developed, or refined. This means that Africa's Gene Revolution is buttressed in important ways by private capital, both in terms of the involvement of small- or medium-scale agro-dealers, billion-dollar multinational corporations, and corporate lobby groups such as the International Service for the Acquisition of Agri-biotech Applications (ISAAA), whose mandate is to connect agricultural biotechnology with farmers in developing countries who lack the funds, skills, and infrastructure to implement it by linking private sector actors from the Global North with public institutions based in the Global South.[52]

At its core, Africa's Gene Revolution is a complex alliance of philanthropic foundations, biotechnology companies, and bilateral development donors who have partnered in order to mobilize genetic modification as a tool to address the productivity gap in African agriculture.[53] The organizations underpinning Africa's Gene Revolution constitute a whirlwind of difficult-to-penetrate partnerships, collaborations, funding arrangements, and acronyms that all share a common commitment to the technological possibilities associated with agricultural biotechnology. This biotech bloc represents an alignment of civil society, state actors, and corporate capital that seeks to advance agricultural biotechnology as the most appropriate mechanism for improving yields in Africa.[54]

Two donors comprise the nucleus of Africa's Gene Revolution: the BMGF and USAID. For decades these two organizations have played an essential role in financing research projects, training students, building infrastructure, supporting regulatory institutions, and constructing promotional and awareness campaigns. USAID's activities have been concentrated in two streams: research and promotion. Its research arm is known as the Agricultural Biotechnology Support Project (ABSP), a consortium of public and private organizations operated via the Collaborative Agricultural Biotechnology Initiative (CABIO) branch. ABSP focused their activities on enabling the commercialization of end products by investing in the full life cycle of product development, including the training of research scientists, expanding domestic biosafety capacity, infrastructure development, and delivery and distribution networks, as well as the design of permissive regulatory structures.[55] Its promotion arm is the Program for Biosafety Systems (PBS), which was established in 2002 to create an enabling environment for biotechnology via lobbying, outreach, and promotion.[56] The BMGF has funnelled substantial dollars to directly supporting specific experimental streams, such as the Water Efficient Maize for Africa (WEMA) project examined in Chapter 6. It further established the African Biosafety Network of Expertise (ABNE) in 2008, with a mission to connect African countries to agricultural biotechnology while supporting training and biosafety regulation.[57]

The biotechnology imperative that fortifies Africa's Green Revolution has also been the subject of criticism by both environmental activists and development scholars. The starting point for many of these critiques is an allegation of continuity: detractors argue that Africa's Green Revolution represents an extension of long-standing attempts by external forces to reshape Africa's agricultural sector.[58] They point out that patterns of land alienation and agricultural reformation have a long history in Africa: critics view this latest paradigm of agricultural development as emulating and propagating many of the key ideals and aspirations that underpinned colonial agricultural development, including the privileging of outside expertise, a partial understanding or misunderstanding of smallholder farmer systems, and an unwavering faith in the power of technology to overcome production constraints.[59]

Scholars in critical agrarian studies object to the ideological foundation of the GR4A, which posits liberalized markets as the sole route for escaping poverty and hunger.[60] They see AGRA as the latest in a long line of attempts by private capital to create a "capitalist farming revolution in Africa" that seeks to dismantle the development state and replace it with a new emphasis on philanthropic capitalism.[61] Within this view, the burgeoning influence and unparalleled scale of the B M G F imbues the G R 4 A with a degree of influence, appetite for risk, and lack of accountability that enables it to remake African smallholder systems into agribusiness value chains that are large scale and capital intensive, and which are then amenable to incorporation into global markets.[62] The result, according to detractors, is that smallholder livelihood strategies become reimagined as barriers to be dismantled, which in turn transforms smallholder farmers into wage labourers displaced to cities.[63]

Opponents further point out the close links between the development donors anchoring this push (such as the B M G F) and the private corporations whose technology is being employed to achieve its goals (such as Monsanto), exposing the cross-pollination of individuals, expertise, and approaches that coalesce around these shared interests. This has led to accusations that G R 4 A is paving the way for biotechnological interests; as one informant told me wryly, "U S A I D is the biggest promoter of its own technology."[64] This accusation of "biotechnology bias" has led some to argue that there are more subtle forces at play, emphasizing the positive public relations for multinational corporations, or a geopolitical angle in which Africa's embrace of G M crops isolates Europe as the only continent taking a more reticent stance against biotechnology.[65] Whether you are for it or against it or somewhere in between, what is undeniable is that Africa is an increasingly important battleground in the global fight over the potential for biotechnology to improve agricultural yields and livelihoods.

WHO AM I?

The first questions to ask whenever someone wades into the waters of the genetic modification debate relate to the person themselves: Who do they work for? What are their political leanings? Who funds their research? My own interest in the debate over G M crops stems from an undergraduate degree in biology, which exposed me to the scientific tools and technologies fuelling the biotech boom. As a master's student I gravitated towards questions that concerned the social and political implications of these new technological innovations. A PhD in geography exposed me to the importance of space and place, which prompted me to go beyond the broad contours of this debate and ask more specific questions, about whether *this particular* G M crop makes sense *in this particular* agricultural context. I lived in

South Africa for a year while completing my PhD research, which evaluated attempts by settlers and scientists to impose cotton as a commodity crop in the eastern part of the country. At that point in time South Africa was the only country on the continent to have commercialized GM technology, and I became interested in whether these highly trumpeted technologies could succeed in boosting yields for some of the country's most vulnerable farmers.

Over the past ten years I have undertaken a total of twelve fieldwork trips across five countries in southern, eastern, and western Africa, using a combination of qualitative and quantitative methods to assess the perspectives and priorities of hundreds of smallholder farmers and amplify their voices on whether GM crops make sense given the conditions they face on the ground. I have further convened numerous workshops and meetings with relevant community leaders, members of parliament, NGOs, and extension officers, and shared this work with media, in print, on radio, on television, and online.

This book is an attempt to scale up the research stemming from this grounded, context-specific inquiry, in order to offer broader assessments of the potential for agricultural biotechnology to achieve its stated goals of mitigating poverty and alleviating hunger. The research that forms the groundwork of this work has been funded by philanthropic foundations and national granting agencies that have no stake in Africa's Gene Revolution. I do not receive funds or support from private companies, nor have I received funds from any group campaigning against biotech. I have published in academic journals favoured by those more supportive of genetic modification (such as *AgBioForum*), as well as journals favoured by critical social scientists (such as the *Journal of Peasant Studies*).

My attempts to occupy the middle ground have not always been well received. In Kenya, some pro-GMO campaigners branded me a fraud and a liar. My 2015 research trip caused such ire that an official within the country's Ministry of Education, Science and Technology warned colleagues that I was on a "mission aimed at throwing controversy around the names of Kenyans known to be supportive of biotechnology."[66] I was subsequently blacklisted from visiting any of the national research centres by the coordinator of the government's biotechnology program. In Ghana, I was dismissed by anti-GMO campaigners who grew frustrated by my tiresome objection to some of their ill-advised and misleading claims. In Uganda, where I conduct the majority of my work, I have taken flack from both sides. A cohort of pro-GMO campaigners have long attempted to discredit my peer-reviewed contributions to the debate, dismissing them as "insulting to us, abusing our intellect, causing un-necessary panic to the public, lacking factual information and written to mislead readers with the intention which to date we are still investigating."[67] The country's most virulent critic of GM crops has urged me to rethink my position and join him in condemning proponents of genetic modification as criminals. I admit that these attacks have, at times, been hurtful. But I take solace in the words of an encouraging mentor, who once reassured me: "if both sides are angry at you, then you can be pretty sure that you're doing something right."

A POLITICAL ECOLOGY
OF AFRICA'S GENE REVOLUTION

My analysis of Africa's Gene Revolution is informed by a range of theoretical perspectives that converge within the broad canopy of political ecology. At its core, political ecology is a conceptual and analytical lens that straddles the social and natural sciences by combining inquiries into political economy and biophysical systems. It is a perspective that is deliberately expansive and ambiguous, encompassing insights from a range of theoretical traditions including Marxist political economy, post-structuralism, science and technology studies, farmer systems theory, and critical agrarian studies. Political ecology's breadth of coverage and pluralistic tendencies are, in my view, its greatest strengths, leading to systematic inquiries of rural development that are inclusive enough to take into account simultaneous processes of environmental change, representational politics, the logic and imperatives of capitalist growth, and networks of production.

My operationalization of political ecology is grounded in four key pillars. First is the inextricability of nature and society. Influenced heavily by Marxist political economy, political ecology problematizes the notion that environmental issues are either natural or social in origin, committed instead to unravelling how biophysical and human processes interact to create outcomes within the human-environment nexus. This insight is particularly valuable in a study focused on agriculture as a driver of rural development. Political ecology's focus on connecting biotic and abiotic components alongside human interventions underlines how natural and social systems interact to produce or mediate rural development outcomes. This emphasis on

understanding natural-social relations syncs with my preferred definition of agronomy, taken from *Larousse Agricole*, which similarly
stresses the importance of connections: "agronomy is the scientific
study of the relationships between cultivated plants, the environment
(soil, climate) and agricultural techniques."[68] Political ecology offers
a useful extension to agronomy's holistic, systems-based approach by
widening the analytical lens to include social and political relations
alongside the biophysical, in order to provide a more comprehensive
assessment of agricultural development outcomes.

Political ecology's emphasis on the inextricability of nature and
society challenges both the exclusivity and the objectivity of scientific
knowledge. Influenced heavily by science and technology studies,
which has exposed science as a socially constructed entity, political
ecology seeks to uncover how human factors shape what counts as
scientific knowledge, and unravel how scientific authority reflects and
refracts a particular set of biases and assumptions.

In the context of Africa's Gene Revolution, the blurring of science
and politics means exposing how scientific credibility – in the form
of individual experts, renowned research centres, state-run experimentation stations, prestigious universities, and multilateral donor
agencies – is created and perpetuated. What is needed is a critical
analysis of the plant breeding efforts that fall under the banner of
Africa's Gene Revolution, one that engages with sticky questions
around the political and economic assumptions underpinning it, as
well as the associated implications of genetic modification's restrictive
growing regimes. This study serves as a rebuke to calls that the debate
over genetic modification needs to be grounded in science as opposed
to politics, arguing instead that all science is political.[69]

Political ecology's second pillar is the recognition that the non-
human environment constitutes a politicized space contested by actors
with varying degrees of power. Dominant interests produce influential
portrayals of environmental debates, each bolstered by a distinct set
of values, goals, and assumptions. Political ecology sets out to "shatter
comfortable and simplistic 'truths' about the relationship between
society and its natural environment."[70] Political ecology's tendency
towards interrogating environmental narratives reveals the profound
influence of post-structuralism and discourse theory, as its adherents
seek to deconstruct and challenge dominant environmental narratives
based on inequities in access, control, and ownership. Placing the
analytical focus on knowledge politics within a study of agricultural

development requires an interrogation of the methods and data that underlie these ventures, including questions first posed by Scoones and Thompson: "What interests frame the dominant narratives driving this policy agenda? What alternatives are excluded as a consequence? ... And what processes of agrarian change are promoted as a result?"[71]

This book seeks to unearth the political and economic interests underpinning Africa's Gene Revolution in order to discover why GM crops are being favoured over other possible pathways of agricultural development. In the narrative that follows, genetic modification is understood as a nexus of social relations that needs to be unpacked to reveal the financial, political, and cultural interests embedded within.[72] Spurred by political ecology's deep roots in Marxist traditions, this manuscript shines a light on the profound ways that private capital shapes Africa's Gene Revolution by interrogating underlying assumptions of modernity (i.e., technological innovation is the best strategy for enhancing agricultural yields) and rationality (i.e., farmers will always prioritize agricultural yields over other growing characteristics).

Political ecology's preoccupation with power prompts further engagement with the final question broached by Scoones and Thompson: who wins and who loses within agricultural development interventions? Distributional dynamics are paramount when analyzing the impacts of any new agricultural innovation. This requires paying attention to how benefits associated with genetic modification are segregated based on categories of power including class, race, age, and gender. Here, I am centrally concerned with boasts of genetic modification as a "pro-poor" technology that can help to empower smallholder farmers, and how such benefits are refracted across these other categories of power.[73]

The third pillar of political ecology is adherence to multiscalar analysis. The entry point here is the recognition that the social, political, economic, and ecological processes that impact the success or failure of development interventions are overlapping and interconnected. Political ecologists tout the value of nested scales as a device for unravelling these layers of explanation; the metaphor used most often here is that of Russian dolls, wherein each scale of analysis is designed to fit neatly into the others. This examination of Africa's Gene Revolution aims to understand how farm-level outcomes are embedded in processes that extend well beyond the farm gate.[74] This requires creating chains of explanation that connect the micropolitics of farmer decision-making with community-level issues of market access, credit availability and extension services, state-led breeding programs, the funding mechanisms of development donors, and the priorities of multilateral institutions.[75]

Political ecology's commitment to multiscalar analysis extends across time as well as space. A study of agricultural innovation for development requires a deep theoretical commitment to incorporating the "historical specificities" of agrarian change, and to appreciate the relevance of precedents dating back to the colonial era.[76] This entails examining the past to learn more about the present, critically analyzing whether and how Africa's Gene Revolution replicates its predecessors' patterns. One final point of emphasis here is that this engagement with multiscalar analysis is at once analytical and geographical. Political ecology recognizes that agricultural innovations will perform differently in different agro-ecological settings, which requires stressing the spatial and social specificities of agrarian change.[77]

This commitment to engaging with scale analytically, temporally, and geographically translates into a relentless emphasis on context. In my view, political ecology's greatest contribution within the realm of agricultural development has been to expose the unintended consequences of well-meaning interventions that were designed and implemented with insufficient understanding of the myriad contextual factors that determine outcomes of agrarian change.[78] No study is perfect, but this one aims to redress these shortcomings by foregrounding the contextualized micro- and regional dynamics of farming systems in order to assess whether GM crops can achieve their lofty aims.

A commitment to employing multiple methodologies constitutes the fourth and final pillar of political ecology. The methodological approach employed most commonly by political ecologists is best characterized as "everything goes." This phrasing reflects political ecology's inclusive attitude as well as a deep commitment to a plurality of methodological approaches: all tools are welcome and the more the better. This stress on triangulation produces studies that systematically incorporate multiple layers of data: political ecologists recognize that each approach produces knowledge that is partial, so the greater the breadth incorporated the more complete the portrayal. Most studies in political ecology remain anchored by qualitative research of some sort, which tends to be beefed up with quantitative methods, document analysis, or studies of land use change. According to Perreault, Bridge, and McCarthy, "In this sense political ecology continues to be radically experimental in the field, mixing conceptual genres and methodological registers in an effort to understand and transform socio-ecological relation."[79]

The methodological approach employed here exemplifies this commitment to pragmatism over pedigree.[80] Over the past ten years I have completed more than 125 interviews with research scientists, government officials, development donors, and NGO activists. I facilitated on-farm ranking exercises with over 200 farmers and analyzed these large data sets using statistical methods. I subsequently convened gender-specific focus groups with over one hundred farmers to uncover the reasoning and rationale behind these aggregate responses. I have undertaken dozens of transect walks (guided tours) with farmers across five African countries. These qualitative and quantitative data sets were further supplemented by extensive document analysis that included relevant reports and media coverage, as well as available secondary data chronicling changes in land size, debt levels, and crop prices. Taken together, this layered approach should help to capture the reverberations of farm-level dynamics across multiple scales and offer a systematic assessment of the potential for GM crops to alleviate poverty and hunger in Africa.

The integration of these four pillars generates a theoretical approach that is open-minded, flexible, and robust. The value of political ecology as a conceptual lens is perhaps best captured by the oppositional notion of apolitical ecology: according to Paul Robbins, the best strategy for demonstrating the value of political ecology is to pinpoint the perils of its absence.[81] Two examples of apolitical ecologies were featured in the preceding pages. The first is the triumphant narrative confidently embraced by development donors, which heralds genetic modification as the most promising solution to Africa's pernicious yield gap. The second is this narrative's direct counterpoint, the critique of GM crops as a nefarious plot being designed by greedy multinational corporations. The pages that follow will expose the fallacy of both extremes. Instead, a political ecology of Africa's Gene Revolution navigates the murky middle ground in order to assess whether GM crops can enhance yields and livelihoods for African farmers.

THIS BOOK'S CONTRIBUTION

Debates around the role of new breeding technologies as tools to alleviate poverty and hunger are, by their very nature, polarized and politicized. This book offers the first data-driven assessment of the potential impact agricultural biotechnology could have in enhancing farming livelihoods across the continent. It builds upon ten years of primary research investigating such issues in various African nations, including research in South Africa (2004–2008), Uganda (2009–2018), and Kenya (2010 and 2015–2018), as well as shorter visits to Ghana and Burkina Faso (2015). The analysis and insights that make up the following pages represent my own views, which are the result of careful examination of data and my own experience conversing with farmers.

Many books have been written problematizing the treatment of Africa as a homogenous, undifferentiated whole. This book navigates this hazard by zeroing in on case studies that are relevant and representative. So, in section 2, examining GM commodity crops designed to enhance insect resistance and herbicide tolerance, I have chosen to focus on the cases in which the empirical record is the longest and most robust: GM cotton in South Africa, Burkina Faso, Sudan, and Uganda, and GM maize in South Africa and Egypt. I omitted the case of GM soybean in South Africa due to its more recent release, its limited geographical exposure, and its marginal importance within smallholder farming systems. Similarly, in section 3, investigating GM crops targeting traits and crops that matter to poor farmers, I have opted to examine the cases of biofortified sweet potato and insect-resistant maize in Kenya because these served as crucial, early precedents, alongside drought-tolerant maize in Kenya and cooking banana in Uganda as cases that exemplify the new wave of GM carbohydrate staple crops. This meant not including a treatment of experimental programs that are not as far along, including virus-resistant cassava, biofortified sorghum, and insect-resistant cowpea, amongst others. In sum, this book offers a partial, but not exhaustive, examination of the countries and crops whose experiences tell us the most about the potential for Africa's Gene Revolution to enhance the lives of smallholder farmers.

The book uses political ecology to shine a light on what I consider to be the three most important considerations that need to be taken into account when evaluating the potential for GM crops to improve agricultural development in Africa. First is the issue of context. Technologies are not inserted into a vacuum. In order to be successful any new agricultural technology needs to be congruent with the ecological, social, and political contexts of the intended adopter. Thus, this book pays particular attention to the "fit" between the GM crops being proposed and the farming system employed, in order to offer a rigorous evaluation of whether this particular technology has the potential to positively impact both yields and livelihoods. Second is the question of how we measure the impact of technological change. As one recent systematic review makes clear, the vast majority of studies offer a very narrow measure of the impacts of GM crops: 84 per cent of studies emphasize economics, leading these authors to conclude that "the literature is dominated by the study of economic impact."[82] This book takes an integrated approach to methods, assessing economic impacts alongside social, ecological, and political considerations. Third is the issue of scale. GM crops are ubiquitous throughout the world, but almost all crops are grown under industrial agriculture: large scale, monoculture, heavily capitalized, and mechanized. The core question of this book is whether GM crops can succeed in a very different model of agriculture that is dominant throughout Africa: one that is decidedly small scale, focused on polyculture, with little access to inputs.

It is equally important to understand what this book is *not* about. This manuscript is primarily focused on the promise GM crops offer to African growers, but remains relatively silent on the potential for African consumers (though it is important to note that this distinction remains very fluid across much of the continent, where the vast majority of Africans double as both producers and consumers of food). As a result, little weight is given to the health dimensions of GM crops; that is, I do not dwell on issues of food safety such as toxicity or allerginity, mostly because, in my mind, these risks are relatively minimal, and because my research emphasizes the social and economic implications of new biotechnologies.

The book tackles these urgent questions by splitting them up into three successive parts. Section 1 introduces two key background elements relevant to this debate: the breeding technologies themselves and the regulations that have emerged to manage their associated risks. Section 2 examines the past fifteen years of GM crops in Africa, which have centred on first-generation GM technologies, consisting of herbicide-tolerant and insect-resistant versions of cotton and maize. Section 3 deals with the new, second-generation GM crops: carbohydrate staple crops that are being engineered under public-private partnerships (P3S) that will be made available to farmers licence-free, designed to respond to the needs and priorities of smallholder farmers. In the conclusion, I bring all of these components together, putting evidence before ideology to assess the potential for these technologies to alleviate poverty and hunger.

SECTION ONE

The Promotion, Science, and Regulation of GM Crops

1

Talking Technology: Plant Breeding for Agricultural Development

One of the most confusing elements of the debate around the potential for GM crops to improve African agricultural development is the complex and at times contradictory presentation of the technology itself. Efforts to categorize the broad range of plant breeding biotechnologies, how they work, and their implications for agricultural development leave many non-scientists feeling intimidated or overwhelmed. This chapter offers a pithy synopsis of the full suite of breeding technologies, from conventional plant breeding and improved hybrids through to agricultural biotechnology and genetic modification. The goal here is not to be exhaustive, but rather to outline the contours of this technological landscape, in order to enable a nuanced evaluation of their associated possibilities for smallholder farmers across Africa.

MODIFYING GENES VIA TRADITIONAL TECHNIQUES (A.K.A. CONVENTIONAL PLANT BREEDING)

The main source of confusion around genetic modification stems from the term itself. After all, many wonder, humans have been modifying genes since we began practicing agriculture more than 10,000 years ago. So aren't all the crops that we grow genetically modified? It is certainly true that human beings have been altering the genetic makeup of plants by selectively breeding for favourable traits since our forbearers first began sedentary cultivation more than 12,000 years ago. Indeed, our ancestors' commitment to this process of seeking out, preserving, and expanding their preferred medley of plants has been one of the most important factors that has shaped our species' history.

Farmers have been the caretakers of plant genetic diversity for millennia, selectively breeding for observable preferential traits such as higher yields, larger fruit, or pest resistance. The most basic form of selective breeding is known as mass selection. Individual plants displaying the desired trait would be bred together in order to increase the trait's frequency and expression. The best performers among this next generation would then be selected for another round of crossing, and the process would continue until a new variety emerged that consistently displayed the desired characteristics across all individuals within the population. Crucially, the success of mass selection was dependent upon the genetic variation within a given plant population, which remains a vital prerequisite for identifying and stabilizing the characteristic of interest.

In the early twentieth century, a more precise form of selective breeding emerged known as backcrossing. Rather than searching for observable traits among the population as a whole, backcrossing offered a more targeted approach for introducing a specific trait to existing varieties. The process begins with two individuals: a donor organism with visible expression of the desired trait, and a recipient organism, often known as an elite or host variety, which demonstrates a number of other favourable traits but lacks the desirable trait expressed in the donor organism. The goal is to interbreed the donor and recipient organism so that its progeny expresses a stable and consistent version of the desired trait, while still maintaining all of the recipient organism's favourable characteristics. This first generation progeny is then selected for expression of the desired trait and crossed again (i.e., backcrossed) with the recipient organism. This process of backcrossing is repeated until the desired trait is expressed reliably and uniformly across all the individuals within a given generation.

The process of selective breeding succeeded in generating agricultural surpluses that sustained the entrenchment and expansion of contemporary civilization. Across sub-Saharan Africa (SSA), both colonial and independent regimes invested heavily in improving yield, quality, and adaptation via selective breeding. In cotton, for instance, colonial ventures such as the Empire Cotton Growing Corporation relied on mass selection to develop improved varieties resistant to local insect pests while simultaneously satisfying the specific desires of British cotton buyers.[1] Selective breeding also notched noteworthy gains in sorghum, where breeding efforts focused on enhancing plant height, maturation times, and resistance to both abiotic and biotic stressors.[2] In cassava, plant breeders used a series of backcrossing techniques to develop heartier varieties that could resist the most pernicious diseases, including cassava mosaic disease and cassava bacterial blight.[3] In sum, selective breeding played a vital role in increasing both the productivity and adaptability of African crops.

But selective breeding also has considerable limitations. First, it is imprecise. Selective breeding amounts to a system of trial and error, in which many genes that code for non-targeted traits are inevitably transferred along the way. This lack of precision is further exacerbated by the fact that selective breeding is based on phenotypic expression – that is, the visible expression of the trait in question – which is subjective and limits the precision of the breeding process. Second, selective breeding is slow and time consuming. Conventional plant breeding techniques often rely on the crossing of hundreds or even thousands of individual plants, which means that producing a line of stable crosses can take anywhere from several months to a few years. Third, selective breeding is inconsistent. Advances in mechanization and irrigation have entrenched the need for synchronized periods of germination, maturation, and harvesting amongst individual plants. At the same time, food tastes have evolved to the point where consumers are demanding standardized agricultural products that express their desired traits in a predictable and uniform manner. These levels of consistency are difficult to achieve within selective breeding. Fourth, and most crucially, selective breeding requires a certain level of genetic variability within a given population, which is undermined by the emphasis on producing standardized individuals. In this way, selective breeding is self-defeating, in that it leads to genetically uniform varieties that destabilize the very genetic diversity upon which it depends.

Plant breeding efforts in the mid-twentieth century focused on overcoming these declines in genetic variability. The most significant advance was the insight that crossing genetically divergent inbred individuals leads to progeny that yield higher and develop faster than either parent, a process known as hybrid vigour, or heterosis. An added advantage is that these first-generation progeny tend to express uniform growing characteristics, including germination rates, stem development, and growth rates, making the crops easier to manage, particularly for a farmer relying on mechanized inputs and irrigation.

These superior varieties, known as hybrids, began being sold commercially in the 1930s. It is difficult to understate their significance in changing global agricultural production. By 1960 almost 98 per cent of the maize grown in the United States was hybrid.[4] Developing countries were slower to adopt the use of hybrids, with penetration only reaching the 60 per cent threshold in the early 1990s.[5] The limitation of the single growing season made the investment in agricultural productivity extremely profitable, sparking a massive splurge on the part of private companies who for the first time had managed to overcome the "reproducibility" problem of agriculture. As Jack Kloppenburg observed, "the natural characteristics of the seed constitute a biological barrier to its commodification," as it can be replanted and recycled without incurring additional cost.[6] But declines in hybrid vigour made it uneconomical for farmers to replant seed, opting instead to buy new seed at the start of every growing season, dramatically accelerating the profitability of the enterprise of creating superior crop varieties. In this way, "hybridization has proved to be an eminently effective technological solution to the biological barrier that historically prevented more than a minimum of private investment in crop improvement."[7] The rapid proliferation of hybrids in wealthier countries led to calls to share these yield advances with poorer farmers, with the resultant high-yielding varieties emerging as the centrepiece of the Green Revolution outlined in the introduction.

Hybrid technology did not spread as quickly in SSA as it did in Latin America and South Asia, but it still played an important role within the continent's recent agricultural history. By the mid-twentieth century, maize breeders in southern Africa had developed a high-yielding hybrid known as SR52, which outyielded the best existing varieties by more than 300 per cent.[8] But achieving this level of heterosis required a longer growing season, heavy use of fertilizers, and the regular repurchasing of seed, as the benefits of hybrid vigour declined sharply after the first generation. This led to a highly uneven maize sector in which larger, wealthier, and predominantly white farmers shifted almost exclusively to hybrids while smaller, poorer, and primarily black farmers relied almost exclusively on their conventional seeds (this asymmetry is discussed in more detail in chapter 4). Hybrids also brought about sizeable jumps in yield in sorghum and cassava, though as with maize, smallholder growers were largely excluded from these yield-enhancing benefits.

The proliferation of hybrid seeds generated huge profits, giving rise to a new class of professional agronomists who displaced farmers as the caretakers of the world's plant genetic resources. Or, in the Marxist lexicon used by Kloppenburg, the rise of hybrid seed precipitated "the historical transformation of the seed from a public good produced and reproduced by farmers into a commodity that is a mechanism for the accumulation and reproduction of capital."[9] Kloppenburg was the first scholar to delineate the full implications of this transformation: plant breeding shifted from a public to a private good, as companies sought to capture the financial benefits of these increasingly remunerative commodities known as seeds. Farming was transformed from a production process that was largely self-sufficient to one in which the tasks of crop cultivation and seed selection were broken up: the former remained in the purview of farmers but the latter was taken over by private corporations. The legacy of conventional plant breeding is reflected in the transformation of thousands of locally adapted landraces into dozens of higher-yielding but genetically uniform varieties. The next section turns to the more recent advances in agricultural biotechnology, which have further accelerated and entrenched this process.

MODIFYING GENES VIA AGRICULTURAL BIOTECHNOLOGY (A.K.A. MODERN PLANT BREEDING)

Agricultural biotechnology refers to a set of technologies that uses molecular understanding of an organism's genome to improve its performance. While conventional plant breeding operates at the scale of individual organisms and relies on sexual reproduction to transfer genetic material, agricultural biotechnology allows breeders to manipulate genes at the cellular level and create new mechanisms for genetic transfer that bypass sexual reproduction. These laboratory techniques enabled scientists to identify genes that code for desirable traits and engineer these genes in ways that enhance their advantageous characteristics. Combining insights from molecular genetics, plant physiology, cell biology, and emerging fields such as bioinformatics, agricultural biotechnology represents a departure from conventional plant breeding in that it allows scientists to move beyond breeding for advantageous traits based on their visible expression (known as phenotypic selection) to breeding for advantageous traits based on genetic expression (known as genotypic selection). Phenotypic selection carries significant limitations: some beneficial traits are inconsistently expressed or triggered by abiotic factors; for example, a trait of drought resistance will be difficult to identify in a year of heavy rains, while a trait of insect resistance will be difficult to identify in a year of low pest incidence. Other characteristics take years to reveal themselves or are expressed inconsistently, adding to the time burden of conventional breeding. Plus, most phenotypic characteristics are dependent on multiple genes, so agronomists must seek to ensure the expression of these beneficial characteristics while not losing the other beneficial traits that are outside the realm of this target trait. Agricultural biotechnology is able to overcome these obstacles by utilizing the gene that codes for the desirable trait as its starting point, offering plant breeders a more precise and more efficient tool in the search for enhanced agronomic attributes.

Marker-Assisted Selection

In the late 1970s scientists first discovered the existence of molecular markers, identifiable DNA sequences located in close proximity to a particular gene that codes for a desired trait. By the 1980s breeders were able to utilize these markers as landmarks to locate target genes, which greatly enhanced both the speed and precision of backcross breeding.[10] Today, marker-assisted selection encompasses a range of techniques related to the identification, tracking, and monitoring of specific regions of the genome. It represents an important step forward in terms of the accuracy of plant breeding, as it allows agronomists to target the specific gene that codes for the specific desirable trait, without inducing any other genetic changes that accompany conventional techniques.

Marker-assisted selection has been used in limited capacities across SSA, due mostly to limitations related to cost, infrastructure, and expertise. Some progress has been made in pearl millet, in which drought resistance was enhanced, and in rice, where marker-assisted selection was used to overcome a series of productivity constraints including improvements to eating and cooking qualities, resistance to pests and diseases, and stress tolerance.[11] The most significant gains have been in cassava, where marker-assisted selection has been used to shave four years off of the breeding cycle in the search for disease- and insect-resistant varieties.[12] Marker-assisted selection has further allowed scientists to introgress desirable traits into Latin American cassava varieties and establish these in SSA, a process that had previously been hampered by the vulnerability of the South American varieties to the pests and diseases they became exposed to as a result of their transplantation.[13]

Tissue Culture

Tissue culture is a form of micro-propagation that allows for the replication, regeneration, and rapid dissemination of planting materials that are genetically identical and disease free. The process begins with the surgical removal of a tiny piece of the plant, which is then submerged in a nutrient-rich liquid designed to promote regrowth.

Tissue culture has figured prominently in agricultural development efforts as a technique for the multiplication of clean planting materials free of pests and disease. It has proven especially advantageous amongst crops that propagate vegetatively such as sweet potato, cassava, and banana. African farmers who cultivate these crops tend to rely on recycled planting materials by cutting or transplanting source materials from standing crops, which facilitates the accumulation and propagation of systemic pathogens that substantially reduce overall yields and quality. Tissue culture is a means of interrupting this negative cycle of pathogen reproduction. In sweet potato, the introduction of healthy planting materials via tissue culture was able to increase yields by more than 30 per cent in Ghana and by more than 500 per cent in Uganda.[14] In cassava, tissue culture has facilitated the importation of new varieties from Latin America, minimizing concerns over the unintentional introduction of new diseases and making the large-scale transfer of germplasms more cost-effective.[15] In banana, tissue culture has played a major role in reviving the Kenyan banana industry: farmers there reported that the reduction in pest and disease damage increased yields, and responded favourably to a shorter maturing period and greater uniformity in bunch size.[16]

Still, the overall penetration of tissue culture in SSA remains limited, due to issues with cost as well as low success rates in hardening plantlets to ensure they survive acclimatization out of the septic cultivation phase once transplanted into farmers' fields. Specifically, major drawbacks that have been identified include the high cost of buying tissue culture plantlets (in Uganda, for instance, such plantlets are often four to six times more expensive than a non-tissue culture equivalent), as well as increased requirements for water, labour, and fertilizer. Existing experimental efforts designed to lower tissue culture costs are underway, but thus far these cost-reducing efforts have come up short.[17]

Genetic Modification

GM crops got their name from the umbrella group of laboratory techniques – known collectively as genetic engineering – that are used to permanently alter an organism's DNA with the aim of enhancing the expression of favourable traits. The starting point for genetic modification is the same as that for the conventional breeding technologies introduced earlier: a donor organism with visible expression of the desired trait, and a recipient organism, often known as an elite or host variety, which demonstrates a number of other favourable traits but lacks the desirable trait expressed in the donor organism. The goal is to interbreed these two organisms to produce a new generation that contains all of the benefits of the recipient organism with the added benefit of the trait conferred from the donor organism. What is distinct about genetic modification is the array of tools that are used to achieve this outcome, the first of which emerged in the 1980s. Known as recombinant DNA techniques, these practices rely on intermediaries to transfer foreign DNA into the recipient organism: the two most common are a gene gun that bombards the cell with heavy metal particles, known as biolistics, and a soil bacterium that serves as a vector. The resultant progeny is identical to its recipient organism parent in every single way, except that it now contains the gene that expresses the beneficial trait that was transplanted from the donor organism (this imported gene is known as a transgene, which is why the process of genetic engineering is also referred to as transgenesis). What is particularly novel about genetic engineering as compared to other plant breeding techniques is that it gives breeders the opportunity to manipulate target genes while leaving the rest of the genome unaffected, and that it allows for the transfer of genes between different species (see figure 1.1).[18]

The process of transformation is often messier in practice than it is in theory. Both methods of transformation, whether by gene gun or bacterial vector, are imprecise and unpredictable. A gene gun can produce multiple versions of an introduced gene, which can lead to erratic and undesired gene expression as well as genetic instability.[19] Critics of the agro-bacterium method suggest that unnecessary

Conventional plant breeding

Genetic modification (transgenesis)

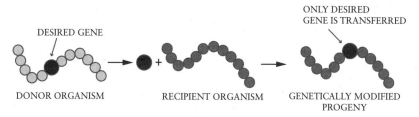

Figure 1.1 Conventional breeding versus genetic modification. Original figure inspired by a diagram at https://www.vox.com/cards/genetically-modified-foods/how-is-gmo-food-different-from-regular-food.

vectors are introduced into the recipient organism whose full implications are unknowable.[20] As with so much of the debate over genetic modification, the implications of this unpredictability associated with the transformation process depends on which side of the debate you are on: biotech boosters reassure us that such negative offshoots can be easily identified and removed during subsequent phases of trait stabilization, while skeptics of genetic modification argue that this high degree of unpredictability is akin to "throwing a grenade in the genome."[21]

For our purpose of examining their potential to enhance agricultural development in ssa, it is most useful to divide genetic modification into three categories.

First-generation GM crops refer to traits and crops designed to make industrial farming more productive and more profitable. The bulk of GM crops planted in the world today belong to one of four global commodities: maize, cotton, soybean, and canola. The overwhelming majority of these four commodity crops are genetically modified to reflect one of two traits, both of which are proprietary technologies owned by the agribusiness giant Monsanto: herbicide resistance, designed to confer tolerance of Monsanto's Roundup Ready glyphosate herbicide, and insect resistance, where genes from a soil bacterium are transplanted to confer resistance to a certain class of insect pest.[22] Today, these four crops and two technologies account for 98 per cent of all GM crops grown around the world.[23] Both are planted throughout SSA, and these cases are discussed in detail in chapters 3 and 4.

Second-generation GM crops, which I have referred to elsewhere as GMO 2.0, are distinct from their first-generation predecessors in a number of ways.[24] Their origins lie not in the private sector but instead in collaboration across multiple sectors. While both herbicide- and insect-resistant GM crops were developed by private seed companies, second-generation GM crops are the result of P3S, facilitated by philanthropic foundations and bilateral development agencies that entice private sector companies to donate their genetic engineering technology licence-free for humanitarian purposes. This has important implications for intellectual property, which are rigidly enforced among first-generation GM crops but are waived for second-generation GM crops to ensure that the resultant technologies are available to poor farmers free of charge.

Second-generation GM crops are explicitly focused on so-called orphan crops – that is, crops that have largely been ignored by private sector investment and innovation. Most of Africa's carbohydrate staples fall within this category of orphan crops, including cassava, sorghum, millet, cowpea, and cooking banana, which together with maize account for more than half of all calories consumed on the continent.[25] It is unsurprising that these crops were left out of the first generation of genetic modification: these are crops that are grown and eaten primarily by the poor. As such, there is little profit potential and no financial incentive for private corporations to invest huge sums in the development of GM versions. There is simply no money to be made from genetically modified versions of African staple crops.

The final feature that distinguishes second-generation GM crops from their first-generation cousins is the traits that are being targeted. The chosen trait within second-generation GM crops is explicitly designed to respond to the needs of smallholder farmers. While 98 per cent of the GM crops being grown today are either herbicide- or insect-resistant, second-generation GM crops are created to respond directly to farmer priorities. Over the past decade more than a dozen experimental programs have been initiated including nutritionally enhanced sorghum, disease-resistant cooking banana, and insect-resistant cowpea. None have reached commercialization, leading Michael Lipton to quip that when it comes to second-generation GM crops the "hype-line" eclipses the actual progress made in the "pipe-line."[26] These new GM technologies and their potential to alleviate poverty and hunger for African small-scale farmers are evaluated in chapters 5, 6, and 7.

Next-Generation GM Technology

One of the most challenging elements of studying the social implications of GM crops is how quickly the science changes. Over the past few years, new techniques have emerged that challenge the three-decades-long distinguishing factor of genetic engineering: its capacity to transfer genetic material from one organism into another, also known as transgenesis. Gene editing is trumpeted as the most promising of these next-generation GM technologies. It allows breeders to edit an organism's genome, either by altering, removing, or turning off a gene that codes for a particular trait. Gene editing is cheaper and often more precise than earlier generations of genetic engineering. Equally enticing for breeders is the fact that gene editing moves beyond some of the most cumbersome debates surrounding earlier generations of genetic engineering. Because it does not involve the transfer of foreign DNA between organisms, gene editing has been largely exempt from the moral and regulatory complexities that have hampered the commercialization of other GM technology.[27] Gene editing is an emerging technology with few practical applications in SSA. Still, preliminary experiments suggest that it can accomplish many of the same goals as earlier generations of genetic engineering, including making plants resistant to pathogens and creating herbicide tolerance.[28]

The two most established platforms for gene editing are zinc finger nucleases (ZFNS) and transcription activator-like effector nucleases (TALENS), both of which target specific locations on a plant's DNA. Widespread use is hampered by issues associated with cost and design specificity, as well as restrictions due to limitations in their applicability for non-conforming genome targets.[29] The most trumpeted mechanism for gene editing is known by the acronym CRISPR (clustered regularly interspaced short palindromic repeats). CRISPR makes use of an enzyme known as Cas9 to identify and edit specific genes of interest, leading to insertions, deletions, or inversions designed to modify phenotypic expression. What is particularly groundbreaking about CRISPR is its accessibility; it radically lowers the cost of gene editing, with the potential to make this technology widely available to agricultural breeding programs without significant infrastructure investment.[30] Because it contains no transgenic residues, it seems likely that CRISPR-edited crops will also be able to circumvent the clunky regulatory measures that have saddled GM crops.[31]

Synthetic biology, derided as "extreme" genetic engineering by its detractors, represents the cutting-edge of agricultural biotechnology. While gene editing alters genes and genetic modification adds foreign DNA, synthetic biology uses computer modelling to build the genetic starting materials. But because this technology is so new, there is little discussion about potential applications or interventions to benefit agricultural development in Africa.

CONCLUSION

As any cell biologist, agronomist, or plant breeder will tell you, the distinctions between conventional techniques (a.k.a. conventional plant breeding) and agricultural biotechnology (a.k.a. modern plant breeding) are much more complex than what is presented here. For example, the above synthesis mentions nothing of induced mutation, a technique that uses artificial catalysts such as X-rays to stimulate genetic mutations, which in turn lead to changes in phenotypic expression.[32] Nor is there any reference to protoplast fusion, which relies on chemical treatments or electric shocks to soften the walls of two different cells and create a nucleus that contains the genetic information from both, making it a form of genetic manipulation that crosses the species barrier but not a form of genetic engineering because it does not employ direct manipulation of the genome.[33]

Similarly, any anti-GMO activist or pro-GMO booster will be miffed by my decision to skirt some of the stickiest questions within the polarized debate over GM crops. Do they lead to negative health impacts? My reading of the empirical evidence suggests there is limited risk here, though I am alarmed by the lack of independent, third-party assessments. Do GM crops represent a threat to the non-human environment? I have more concerns here, especially as they relate to open pollinated varieties, and these will emerge in the chapters that follow. Are GM crops substantially different than conventionally bred crops? As Ronald Herring notes, this is a philosophical question more than a scientific one, dividing those who take an "organismic" view of nature and object to the direct manipulation of the genome as disturbing the natural order by "playing God" from those who take a more "molecularist" view that all life is composed of the same genetic material. Within this view, genetic modification is nothing more than the latest in a long series of technological innovations that humans have used to facilitate the expansion of plants that express favourable traits.[34]

Still, while the all-too-short review offered here is neither comprehensive nor exhaustive, it should be enough to impart sufficient technological background to enable a considered analysis of the potential for GM crops to help smallholder African farmers. When considered within this long view, the evolution of the plant breeding techniques offered here is best understood as a long-term battle between the desire the create high-performing, genetically conforming varieties and the concomitant necessity of maintaining a broad base of genetic variability. For proponents, genetic modification is just another tool at the breeder's disposal to increase genetic variability in search for more frequent and more pronounced expression of beneficial characteristics. For critics, the transition from conventional plant breeding to agricultural biotechnology such as genetic modification is an attempt to "eliminate the barriers to the penetration of agriculture by capital," which has resulted in the loss of farmer control over the means of production, seeds, which are now available as commodities for purchase.[35]

What both sides agree on is that each new plant-breeding technology presents its own particular benefits and risks. The decision about whether to adopt these improved varieties often depend more on economic, political, and social variables than it does on scientific ones. It is to these contested terrains that we now turn.

2

Rules and Regulations: Governing GM Crops in Africa

Regulation remains one of the longest-standing and most contentious elements within the debate over Africa's Gene Revolution. Every new technology presents new risks; regulations emerge as formal mechanisms designed to prevent social harms such as deleterious health effects or negative environmental externalities.[1] Regulations pertaining to GM crops and other forms of agricultural biotechnology are generally grouped together under the term "biosafety," a contraction of "biological safety," which refers to "the avoidance of risk to human health and safety, and to the conservation of the environment, as a result of the use for research and commerce of infectious or genetically modified organisms."[2] International standards of biosafety stem from the Cartagena Protocol on Biosafety, an amendment to the Convention on Biological Diversity that specifies requirements for the safe handling and transport of all living modified organisms, including GM crops. More than fifty African countries have ratified the Protocol since it was opened for signature in 2000. Over the past fifteen years, the focus on biosafety regulations has shifted to the domestic level as individual countries have been tasked with creating a national biosafety framework that would be responsible for managing these potential risks in line with the expectations laid out in the Protocol.[3]

The Cartagena Protocol recognizes national capacity as a prerequisite for the implementation of its mandated measures. To bolster capacity in African countries, development donors initiated programs designed to target training in the areas of risk management, decision-making, and monitoring and evaluation, in order to assist them in developing the necessary technical expertise and infrastructure. While

proponents have sought to portray this debate over rules and regulations as a technical issue related to capacity building, the politics
around genetic modification invariably frame these conversations in
important ways. Decisions around the most contentious items within
each nation's biosafety regime – questions around appropriate legal
authority, liability and redress, risk assessment, transparency, and
participation – are crucial to shaping an individual country's technological trajectory.

This chapter offers an overview of the varying regulatory responses
that have played out across the African continent by grouping them
into four categories emblematic of broader trends: the early adopters
(South Africa, Egypt), the emerging adopters (Uganda, Ghana), the
resisters (Zambia, Zimbabwe, Tanzania), and the renegades (Kenya,
Sudan) (see figure 2.1). My central argument here is that biosafety
regulations across Africa have served a dual function: to mitigate
against any negative health or environmental risk that might result
(its stated purpose) and, simultaneously, to accelerate the expansion
of GM crops across the continent (its actual purpose).

THE EARLY ADOPTERS

South Africa was the first country on the continent to recognize the
potential opportunities and challenges presented by novel biotechnology, establishing the South African Committee for Genetic
Experimentation (SAGENE) in 1978. Originally established to ensure
compatibility with rapidly changing standards coming out of the
United States, SAGENE played a decisive role in awareness raising
and training throughout the 1980s. In 1990 an American seed company requested permission to initiate trials with Bt cotton. Subsequently,
SAGENE was granted statutory accreditation and empowered as the
agency to manage these initial forays.[4] SAGENE emerged as the clear-
inghouse for the slew of experiments on genetic modification that
followed in the 1990s: according to Ayele, its guidelines were so
important they became known as the "green bible."[5]

In 1994 SAGENE was tasked with drafting legislation that would
govern experimentation and commercialization of GM crops. The
GMO Act – originally passed in 1997, and subsequently amended to
ensure compliance with the Cartagena Protocol in 2006 – established
two regulatory bodies: the Executive Council, composed of senior
government representatives and which serves as the decision-making

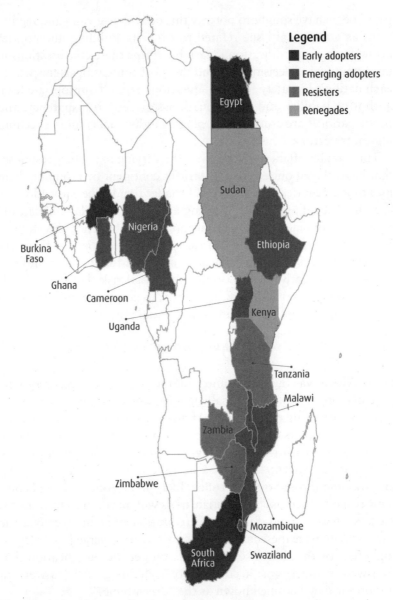

Legend
- ■ Early adopters
- ■ Emerging adopters
- ■ Resisters
- ■ Renegades

Egypt

Sudan

Nigeria

Burkina
Faso

Ghana

Cameroon

Uganda

Ethiopia

Kenya

Tanzania

Malawi

Zambia

Zimbabwe

Mozambique

South
Africa

Swaziland

Figure 2.1 Map showing distribution of four regulatory approaches towards
GM crops. Original figure.

body, and the Advisory Council, made up of independent scientists and which is responsible for undertaking risk assessments and advising the Executive Council. At its core, the GMO Act established a sequential permit system for vetting any new GM technology to ensure it presented no unacceptable levels of risk to human health or the environment. A complementary arm of the South African regulatory framework, the National Biotechnology Strategy, was enacted in 2001 as a means of bolstering secondary considerations such as capacity building, legal complexities, and funding.

The South African regulatory system was the first of its kind and positioned the country as a "biotechnology leader in Africa."[6] Its pioneering regulations were viewed as progressive and enabling by biotech boosters. According to Rijssen et al., a key strength of the governance framework is the division of risk assessment (the Advisory Council) and risk management (the Executive Council). This division of roles ensures that scientific experts are relied upon for advice and guidance but not entrusted with decision-making capabilities, in keeping with international Codex guidelines.[7]

The South African framework was also subject to critique. First, critics stressed the limited role for public participation and the lack of transparency.[8] Others explored how the governance structure privileged the interests of those actors who created it, suggesting that its insular structure allowed scientific issues to overshadow concerns around socio-economics or biodiversity. Still other commentators such as Feris have underlined the limitations of the liability embedded in the South African approach, which places the responsibility for any damage that occurs as a result of genetic modification on the end users rather than the technology developers.[9] Despite these critiques, South Africa's governance framework was the first of its kind on the continent, and was enormously influential in shaping those that came after it.

Egypt is another country that was at the forefront of creating a regulatory system that could facilitate the experimentation and release of GM crops. According to Ayele and Wield, enthusiasm at the prospects of using GM technology in Egypt swelled in the 1980s, propelled by concerns over a rising population and concomitant demands on the domestic food supply, as well as a desire to keep pace with global trends in agricultural intensification.[10] This commitment to "excellence in biotechnology and genetic engineering" led to the establishment of the Agricultural Genetic Engineering Research Institute

(AGERI) in 1990, which became the hub for biotech research in the country.[11] In the early 1990s, AGERI partnered with USAID's ABSP project to increase capacity and infrastructure and "develop the country's agricultural biotechnology system by addressing specific commodity constraints and policy issues including biosafety and intellectual property rights."[12] This included the establishment of the Egyptian Biotechnology Information Centre, which was established as a co-venture with ISAAA, and gave Egypt access to the vast web of training opportunities, workshops, international travel, and communication tools. In total, USAID committed over US$60 million in direct support for the commercialization of GM products, and a further US$600 million in indirect investment towards policy reform, technical assistance, and institution building.[13]

The early 1990s were characterized by a heavy reliance on partnerships with private sector and bilateral donors to increase capacity building (training of scientists abroad), enhance infrastructure (construction of laboratories and greenhouses), and fund ongoing research projects. AGERI's mandate was distinct from those guiding other national research organizations across the continent in that it was explicitly committed to pursuing the commercialization of public research; indeed, part of its operating expenses were designed to be funded from the royalty-based revenues on these commercial applications. AGERI was also exceptional within the domestic landscape: it was officially classified as a "special unit," leaving it outside the control of any one ministry. AGERI was able to benefit from this autonomy, engaging directly in partnerships with donors and contracts with the private sector, as well as fee-for-service offerings in the areas of professional training and DNA sequencing.[14]

AGERI's avant-garde model led it to seek out partnerships with the private sector based on a co-development approach. The former director of AGERI makes the nuanced distinction between this novel approach and the more common vision of technology transfer, in which the developing country partner makes a sizeable contribution to the final output, rather than passively receiving an existing technology from the private sector.[15] In this case, AGERI had already isolated a number of lines of *Bacillus thuringiensis* (Bt) that had pesticidal activity and had its own team of trained scientists and state-of-the-art biocontainment facility. AGERI approached Pioneer Hi-Bred, a US agricultural firm, and initiated a partnership that would allow a two-way sharing of knowledge and technology: Pioneer Hi-Bred offered its technical and legal expertise and its marketing know-how, and committed to train AGERI scientists in new methods of agricultural biotechnology. In turn, AGERI shared its patented Bt proteins and genes, with the aim of developing a commercialized version of Bt maize that could be released to Egyptian farmers within a few years.[16]

While this co-development model yielded only one concrete product, AGERI found more success with an internally focused model that it touted as "All Egyptian." Early on, AGERI made a strategic decision to focus on commercializable GM products that would improve locally adapted crops as opposed to imported products destined for export markets: "In this way, the public's introduction to biotechnology would be in the form of preferred local varieties engineered to overcome local disease or pest problems – products developed at home to benefit Egyptian farmers, growers, and consumers."[17] AGERI initiated a range of projects that used genetic modification to improve varieties of potato, fava bean, squash, and tomato. Research efforts focused primarily on traits such as stress tolerance and pest resistance, in order to produce GM varieties that were "tailored for local conditions and consumer preferences."[18] The two most promising varieties to emerge from these experiments were a transgenic potato designed to resist infestation by the potato tuber moth, and a squash resistant to the damaging zucchini yellow mosaic virus.[19] The late 1990s were a boom time for experimentation on genetic modification in Egypt, with over two dozen separate experiments undertaken between 1995 and 1999.

Egypt established a National Biosafety Committee (NBC) in 1995 and procedures for commercial release in 1998 to regulate these promising experimental ventures. But unlike South Africa, these measures were established not via a national legislative process but rather by ministerial decree. Egypt went on to sign the Cartagena Protocol in 2000 and integrated these principles into its regulatory structure.[20] Egypt's regulatory protocols differed from South Africa's in substance as well as in process. One notable example was Egypt's strategic decision to prioritize the internal research and development of crops of local importance over the import of foreign-owned technologies, which reflected previous investment in pest-resistant versions of potato and zucchini. But the proposed regulations were short on specifics. The former head of AGERI itself criticized these regulations for being overly "general in nature," offering only "brief, uneven lists of recommended practices."[21] Subsequent additions, such as the ministerial decree for the commercial release of GM crops that was published in 1998, were not widely communicated outside of AGERI, leaving key organizations, institutes, and individuals in the dark regarding proper policy and protocols. Predictably, this led to an uneven application of proper procedure: the former head of AGERI has reported instances where institutions were experimenting with GM crops without informing the central authority of their activities, while others proceeded with plans for commercial release without convening the required internal bodies. As a result, critical issues were left unaddressed and underreported; for example, relevant biosafety issues for Bt-resistant crops were ignored, including potential outcrossing between transgenic and wild relatives and the emergence of target pest resistance.[22]

Most critics argued that this inconsistent and incoherent national regulatory approach, applied without full participation of relevant stakeholders, was the major stumbling block in the development of Egypt's biotechnology program.[23] Some stakeholders contended the major obstacle was political, suggesting reluctance on the part of the Egyptian government to counter a negative media campaign launched by opponents,[24] or that Egypt was reluctant to embrace genetic modification for fear of losing out on European markets.[25] Still others argued that the decision to prioritize its own internally developed technologies over more established, existing technologies left the regime with insufficient funds and political capital to build a long-lasting and effective system.[26] In the end, all these forces combined to undermine Egypt's early progress with GM crops.

THE EMERGING ADOPTERS

At the turn of the millennium, South Africa and Egypt were the only two countries in Africa with national regulatory frameworks and functioning experimental programs. The following fifteen years were critical for the expansion of regulatory regimes across the continent: today more than a dozen nations have passed some form of legislation that allows for the experimentation and/or commercialization of GM crops. The path followed by these emerging adopters was distinct from that of the early adopters in a number of ways. Their regulatory regimes were driven largely from the outside in, they followed a more standardized blueprint, and the overarching focus was on regional and continental harmonization more than it was on autonomous, domestic frameworks.

The genesis of the biosafety process for this next generation of African nations can be traced back to the United Nations Environment Programme's Global Environment Facility project (UNEP-GEF), which spanned from 1998 to 2005, including both an initial preparatory phase and an implementation phase. UNEP-GEF provided financial, logistical, and technical support to create formal policies, procedures, and guidelines to help less developed countries conform to the regulations of the Cartagena Protocol. Initiated in eighteen nations across Africa, UNEP-GEF focused on empowering individual countries to build a comprehensive biosafety regime consisting of five pillars: a national government policy on biosafety, a regulatory regime, risk assessment in the form of monitoring and evaluation, capacity building, and a system for ensuring transparency and public participation. A major thrust of the program focused on regulatory harmonization, fuelled by the transboundary nature of GM crops, and individual country representatives were encouraged to "work very closely" with their counterparts elsewhere on the continent.[27] The UNEP-GEF model revolved around creating "a single template" of regulations that was then tailored to each individual country's particular circumstances.[28]

In most of these eighteen countries the UNEP-GEF process provided the regulatory building blocks necessary to manage risks associated with GM crops, including a biotechnology policy, a more specific biosafety bill, and an even more specific set of guidelines and protocols for how these goals would be achieved in practice. For most, the UNEP-GEF process cemented an open regulatory approach to genetic modification that would ensure their easy movement in and out of the country.

The UNEP-GEF process was retired in 2005, leaving a vacuum. While most of the eighteen countries had completed drafts of essential documents by this point, significant investment in infrastructure, capacity building, and expertise was needed in order to convince national governments to adopt these regulations and create the necessary institutions and processes to ensure their implementation. Two donors stepped into to fill this gap. The first was USAID, which sought to ensure the continued investment in permissive regulatory frameworks via its Collaborative Agricultural Biotechnology Initiative (CABIO). CABIO initiated two programs that together shaped the trajectory of regulatory efforts across the continent. The first of these was the Agricultural Biotechnology Support Project (ABSP), currently in its second stage. Established in 1991 and then reconstituted in 2003 as the research arm of CABIO, ABSP II is a consortium of private and public organizations designed to promote biotechnology in three regional centres: South Asia, Southeast Asia, and East Africa. Focused on research programs with clear commercial outcomes, ABSP funds infrastructure, trains research scientists, and directs research into agricultural biotechnology. The second arm of CABIO is the Program for Biosafety Systems (PBS), a program managed by the International Food Policy Research Institute (IFPRI) focused primarily on outreach. Active in ten countries throughout Africa, PBS's mandate involves supporting policy development, capacity building, risk assessment, and regulatory approval in order to create an enabling environment for advancing the safe use of biotechnology into new markets.[29]

The BMGF is the second major donor that stepped in to help fill the gap left by UNEP-GEF. The BMGF invested in a series of platforms designed to accelerate the penetration of new technologies into Africa. Efforts targeting regulatory structures include its investment in Open Forum on Agricultural Biotechnology (OFAB) meetings, which brought together targeted key stakeholders to share information, solidify networks, and coordinate their efforts, and which now operate in eight different countries. The BMGF also joined forces with the New Economic Partnership for Africa's Development initiative (NEPAD) to create the African Biosafety Network of Expertise (ABNE), which was convened in 2010 to "to enhance the capacity of African countries to build functional biosafety regulatory systems."[30] ABNE accomplishes this by offering training programs, providing support for capacity building efforts, and "sensitizing" key government ministers on the potential benefits of GM crops.[31]

The influence of these USAID- and BMGF-funded programs is particularly pronounced in Uganda, a country that boasts one of the largest experimental programs dedicated to agricultural biotechnology in Africa. Uganda was in many ways a typical product of the UNEP-GEF process: by the time this project wrapped up in 2005 it had produced draft versions of a biotechnology bill, a national biotechnology policy, and recommended guidelines for best practices. After UNEP-GEF wrapped up, PBS and ABSP stepped in and played the crucial role of "domesticating" these more general templates to the specific scientific and political realities of Uganda (this according to one of two officials responsible for the UNEP-GEF process in Uganda).[32]

ABSP stepped in as the bridge to ensure that domestic regulators were able to continue with the work begun under the UNEP-GEF process by directing funds to set up and run the biosafety office at the National Council for Science and Technology, the domestic organ charged with overseeing the nation's regulatory protocols.[33] ABSP spearheaded the negotiation around access to Monsanto's Bt cotton technology, providing funding and support for trials at national experimental stations.[34] It also played an active role in supporting infrastructure development, having been one of the major funders for the refurbishment of the National Agricultural Research Organisation's (NARO) Kawanda Agricultural Research Institute, the headquarters for the experimental program into GM banana. ABSP further invested heavily in human capital. It facilitated many of the fully funded scholarships for Ugandan PhD students to study abroad, recruiting these students to work as NARO research scientists once their studies were complete. In a number of instances, ABSP II provided bridging salaries for as many as four years before NARO was able to pick up these salaries in full.[35]

PBS's activities were focused on entrenching the policy instruments that emerged out of the UNEP-GEF process. In Uganda, this meant a primary focus on the government adopting the biosafety bill, a prerequisite for a fully functional biotech system. The original draft languished for years without being formally presented before parliament. Beginning in 2007, PBS oriented its main efforts towards passing the bill. Using a strategy known as net mapping, PBS officials sought to identify the most influential stakeholders in getting the bill passed.[36] The two primary targets that emerged were the prime minister's office, which coordinates the activities of the different regulatory ministries, and the solicitor general's office, which was responsible for drafting the bill. PBS then engaged in a range of strategies designed to influence these offices, either by coordinating workshops, arranging meetings with officials, or inviting representatives to visit confined field trials (CFTs). One of its most successful strategies was its "seeing is believing" tours, which allowed key stakeholders to visit early adopters such as South Africa and Burkina Faso to witness first-hand the success of those countries that had already commercialized GM technology.[37]

The result of these USAID- and BMGF-directed efforts in Uganda was a governance system positioned in a precarious dual role: one that is, simultaneously, charged with regulating and facilitating genetic modification. One member of the country's NBC explains this tension as follows: "when we were initiated we were told, we are not there to block research – so, we are there to promote research, but research that is socially responsible … this is the principle we work from."[38] The danger here is the tension between the economic argument in favour of GM crops as a driver of agricultural biotechnology and the legal obligation to effectively regulate this technology. As I have argued elsewhere, Ugandan regulators are tasked with making influential decisions around health and environmental safety while simultaneously championing the potential benefits of genetic modification.[39] In other words, it is those directly involved in the technology's success who are also evaluating its risks. As a result, it is impossible to determine where promotion ends and regulation begins.

And yet in spite of these coordinated and calculated efforts, proponents have been unable to enact a legislative framework to further their position. The bill that originated from the UNEP-GEF process is quite permissive, with regard to liability (which is fault-based), the burden of socio-economic criteria required before approval (which is unspecified), a lack of any mention of the precautionary principle, and an approach to risk assessment that focuses on the product rather than the process of genetic modification. And yet it remained mired in the approval process for over a decade: it had multiple readings before cabinet, its wording was finalized by the solicitor general's office, and it had multiple readings before parliament. PBS and ABNE pushed to get the bill passed prior to the dissolution of parliament in Februrary 2016. The Biosafety Bill appeared on the order paper during the weeklong rush to pass outstanding legislations prior to the dissolution of parliament, and the minister responsible was even prompted by the speaker to see whether he wanted to put it to a vote. But the minister chose to focus on alternate matters, and the process stalled.[40] After years of continued lobbying and sensitization the bill was finally debated before parliament in December 2017 and ultimately passed. But President Yoweri Museveni refused to give his presidential assent, required to turn the bill into law, citing concerns around patent rights for farmers, preserving the genetic integrity of indigenous varieties, and sanctions for scientists who violate these regulations.[41] One year later, in December 2018, parliament approved a much more restrictive version of the original bill, now titled the Genetic Engineering Regulatory Bill, which includes a strict liability clause that holds the technology developer responsible for any adverse effects that might result, and still-yet-unknown requirements regarding isolation distances and labelling.[42] More than ten years after the original drafting within the UNEP-GEF process, the country's biosafety future remains uncertain.

Ghana is a second country that has emerged as a continental leader in GM crops, due largely to the influence of outside donors. The early history of Ghana's regulatory process mirrors that of Uganda's. Like Uganda, Ghana was a participant in the original UNEP-GEF process, which catalyzed the creation of the country's national biosafety framework in 2004.[43] Like Uganda, USAID's PBS played a key role in "facilitating the establishment of policies and regulations that enable the testing and use of approved bioengineered crops."[44] And, like Uganda, Ghana's Biosafety Act stalled for years in spite of these efforts. To overcome this delay, policy makers tacked on a legislative instrument (LI 1887) to the existing Council for Scientific and Industrial Research Act, which facilitated the implementation of the national biosafety regulations in 2007 and the implementation of the NBC, which oversaw the country's preliminary foray into laboratory experimentation and CFTs.[45] The Biosafety Act, no. 831, was finally passed in 2011, leading to the establishment of the National Biosafety Authority (NBA) as the focal point for all experimentation and eventual commercialization of GM crops.

Three separate experimental trials have received approval and are currently underway: insect-resistant cotton, nitrogen-fixing rice, and insect-resistant cowpea. The first of these is the most interesting from a governance perspective: Ghana became the first country on the continent to forgo single-site CFTs, opting instead to begin directly with no-boundary, multi-locational trials of Bt cotton in six separate growing areas in the northern part of the country. The idea of fast-tracking Bt cotton past the CFT stage was borne out of a meeting in 2013, jointly convened by PBS and ABNE. Two experts were flown in to help facilitate this deliberation: one from PBS and one former member of SAGENE, ABSP II, and UNEP-GEF, who reviewed existing regulatory data from Burkina Faso – Ghana's northern neighbour, who had been growing Bt cotton for over a decade – arguing that these data could be transplanted easily to Ghana, thus alleviating the need for independent trials.[46] The Ghanaian regulators came away convinced: the Ghanaian NBC decided to take advantage of Article 20 of the National Biosafety Act, which states that "The board may exempt a genetically modified organism from certain requirements of section 11, 12, or 13, where it is satisfied that sufficient experience or information that exists to conclude that the genetically modified organism or activity does not pose a significant threat to the environment."

The Ghanaian case set an important precedent: it was the first time an African nation had relied on data from a neighbouring country and proceeded directly to multi-locational trials without first engaging in CFTs. According to ABNE, "This particular trial was special because it was a precedent in the sub-region, being the first time a nation moved straight to multi-location trials using confined field trial data from a neighbouring country with similar agro-ecology."[47] Another biotech booster praised this move as "a significant and bold statement by Ghana."[48] Indeed, the strategy was so successful from ABNE's point of view that it became the template for convincing other African nations to similarly skip the CFT stage: ABNE is undertaking a similar push in Malawi with the express purpose of expediting commercialization of Bt cotton.[49]

The expansion of genetic modification into Ghana stalled briefly due to legal wrangling. Despite the passage of the Act in 2011 the NBA was not established until 2015, due to a combination of limited capacity, a shortage of funds, and misunderstandings on the part of officials charged with implementing the Act's recommendations.[50] During this period the experimentation program continued to be overseen by the NBC, whose role was supposed to be taken over by the NBA in accordance with the 2011 Act.

This oversight resulted in a lawsuit. In February of 2015, Food Sovereignty Ghana (FSG), an NGO composed mostly young university graduates with links to the Rastafarian movement, initiated a legal challenge to the Biosafety Bill at the Human Rights Division in the High Court of Justice, arguing that the NBC had been illegally carrying out the duties of the NBA since the law was inaugurated in 2011. They sought an immediate cessation of plans for commercialization, arguing that this oversight invalidated all of the experimental approvals issued by the NBC since 2011.[51] The list on both sides of this case quickly ballooned: on the side of the complainants, FSG was joined by one of the country's leading political parties (the Convention People's Party), the Vegetarian Association of Ghana, and a farmer cooperative, while the original defendants of the NBC and the Ministry of Agriculture were expanded to include the newly formed NBA (established a mere sixteen days after the launch of the complaint), the attorney-general, and the Ghana National Association of Farmers and Fishermen. Ultimately, the case was dismissed, with the High Court ruling that the applicants' concerns were sufficiently addressed within the country's Biosafety Bill: "The commercialisation of GMOS in Ghana will not in any way affect Ghanaians and the applicant negatively."[52]

THE RESISTERS

A number of African countries have taken a more reticent approach towards GM crops than either of these first two categories. The origins of this reticent approach can be traced back to the first continental-wide effort to address issues of farmers' and community rights over plant genetic resources. Throughout the 1990s a series of meetings were convened bringing together regulators from the Organization of African Union (OAU) and international environmental NGOS. These meetings convinced OAU regulators that member states needed to develop a common position to guide future discussions around ownership and access to genetic resources, which became the African Model Law.[53]

Approved by the then-OAU now-AU (African Union) in 2001, the African Model Law espoused the precautionary principle and laid out an approach that was largely skeptical of new genetic technologies. It prescribed rigorous provisions for risk assessment, including high burdens of proof and a broad scope that spanned the full range of social and economic considerations,[54] as well as robust measures for ensuring public participation and access to information, as well as labelling and traceability. The model law also included comprehensive measures to screen imports, assigned liability to the purveyor of the technology, and most dramatically, placed restrictions on patents over life forms.[55] It was, in sum, a regulatory framework that offered a cautionary approach towards the expansion of new genetic technologies across the continent.

The biotech bloc's reaction to the provisions laid out in the model law was swift and brutal. They dismissed it as restrictive and excessively strict, decrying the influence of European NGOS who had convinced AU officials that GM crops were "a foreign technology imposed upon the people of Africa."[56] Critics denounced the proposed safety standards, which required all experimental trials to be undertaken in complete containment, as excessive and unrealistic, suggesting they served only to undermine domestic experimental programs and "stifle any work that could be done locally."[57] They were especially outraged by the model law's broad regulatory reach and mandatory labelling, which extended not just to genetic modification but also to products derived from genetic modification, arguing that such a restrictive regulatory approach would completely undermine existing efforts towards commercialization: "it can be unequivocally stated that the adoption of this strict AML [African Model Law] will do little to stimulate the adoption of GEOS [genetically engineered organisms] or the development of modern biotechnology within the countries concerned."[58] Muffy Koch, one of the continent's most prominent biotech boosters, described the model law as "a poor working model, designed to impede rather than promote safe and useful technology. Countries choosing this model will end up with regulations that cannot be implemented and will effectively ban the use of all existing products of GM."[59]

Nation states had divergent reactions to the model law. The early and emerging adopters dismissed it as an overreaction, choosing to ignore its conservative approach in favour of the UNEP-GEF's more permissive framework that served to accelerate the spread of GM crops. In countries such as Uganda and Ghana, these regulations were dismissed as "restrictive"[60] and having "zero" influence in shaping domestic regulatory pathways.[61] But this more precautionary approach resonated with other countries that were similarly hesitant about genetic modification.

The most vocal of these was Zambia. Zambia was thrust into the global spotlight fifteen years ago while it was facing one of the worst droughts southern Africa has ever seen, brought upon by a combination of erratic weather, poor governance, and the fallout from a decade of forced economic liberalization measures. The crisis worsened throughout 2002, triggering the importation of food aid by the World Food Programme (WFP) in July of that year. WFP imported over 500,000 tons of food aid, sourced almost entirely from the US, which was distributed among six southern African countries. The revelation that over 75 per cent of this food aid was GM led a number of these countries to reject it, raising concerns about the potential health impacts, the impacts on agricultural biodiversity, and the impact on their ability to export their own crops to Europe in the future.[62] Zambia was the most strident in their opposition. The government convened an emergency meeting of government officials, academic, civil society, and donors, who deliberated on the safety of importing

GM food aid for over six hours. According to Bowman, the official conclusion was one of skepticism around the potential negative environmental and health implications that could result from this import, and concern over the country's capacity to manage any adverse effects.[63] The outcome was a call for an independent, third-party assessment of these potential hazards, suggesting that officials considered the existing assurances to be influenced by institutions and organizations vested in the expansion of genetic modification. A month later, a group of Zambian scientists were invited on a fact-finding mission to the US and Europe, funded by USAID. But this trip only solidified concerns expressed at the earlier meeting: the delegation concluded that the risks of importing GM maize were considerably higher in Zambia, given that maize was their staple crop. These scientists further concluded that US promoters and regulators were too close to the biotechnology industry and were treating them as if they were uneducated.[64]

The US reaction was swift and cutting, accusing the Zambian regulators of being "ignorant" and "uninformed." A delegation from the US Congress visited the country in the midst of this controversy and declared that, "if Zambian policy makers rejected it, they had not grasped the science."[65] Compromise propositions put forward by six southern African nations, including shifting towards in-cash versus in-kind food aid or milling the maize before sending it (which would ensure that the grains could not be replanted), were rejected by the US. In the end, South Africa agreed to mill the American maize, and the five other African countries subsequently agreed to its import. Zambia was the final holdout and refused the import of any GM maize until a team of Zambian scientists could study the impacts. This study was completed in 2004, concluding the risks of importing milled GM maize were still too high, which led the government to formally reject the GM food aid offered more than two years before.

Zambia's stubborn opposition during this time of agricultural crisis earned the country its reputation as the African country most resistant to GM crops. It formalized this stance via its Biosafety Policy (2003) and Biosafety Act (2007), both of which offer a hard-line approach to governance of genetic modification, rooted firmly in the principles articulated in the African Model Law. This legislative framework recognizes the precautionary principle as a guiding principle, mandates rigorous risk assessment and public participation in all decision-making, and includes a liability clause that places responsibility for any negative outcomes on the technology purvey-ors.[66] In addition, Zambia is the only country on the continent to construct a GM detection lab, which was funded by the government of Norway in 2005. This allows for a continual process of testing and evaluation of crops along its border to detect whether any GM content is crossing its borders.[67]

While Zambia remains the symbol for homegrown opposition on the African continent, there are indications that its position has softened considerably over the past ten years. The Zambia National Farmers' Union – one of the most ardent supporters of the original ban – has indicated its interest in reviving the debate, based on new technological developments.[68] The succession of a new president in 2008 signalled another important shift, though the new president's mellower stance has conflicted with other hard-liners within the cabinet. Recently, there has been increasing pressure from South African growers and retailers to be allowed entry into Zambia's domestic market, a move that was approved by the country's NBA in 2016.[69]

If Zambia's opposition to GM crops is now beginning to soften, then Zimbabwe seems poised to overtake it as the continent's most staunch opponent of GM crops. Zimbabwe was one of the six coun-tries that joined Zambia in initially refusing GM food aid back in 2002. Their concerns revolved primarily around potential environ-mental dangers, and they were convinced by evidence linking GM crops with negative health outcomes.[70] Eventually, Zimbabwe agreed to import GM maize that had been milled in South Africa, though they retained a legislative framework that was on the more onerous end of the spectrum, particularly in the area of risk assessment.

The debate over GM crops has recently picked up in Zimbabwe, echoing the debate that took place back in 2002. Southern Africa is currently embroiled in a terrible drought that has left more than 1.5 million people food insecure.[71] The situation is particularly dire in Zimbabwe, where a recent report estimated that over 60 per cent of the population will require food aid. This puts the country in a precarious position, as its most consistent sources of food aid – Tanzania and Zambia – are also suffering from low harvests and have prioritized their own domestic production efforts.[72] In February 2016 the government launched a US$1.5 billion drought relief appeal to overcome its domestic food shortage. Despite these dire circumstances, Zimbabwe's agricultural minister remains adamant that his government will continue to refuse GM food aid: "When it comes to GMOs you know the position of government is very clear, we do not accept GMOS as we are protecting the environment from the grain point of view ... So far, there are places where one can obtain non-GMO grain in Zambia, South Africa, as well as in the Ukraine."[73] He also said the government will monitor and inspect imports entering the country to ensure they do not include any GM content.

This reluctant approach has received hefty international criticism. Much of this opposition has coalesced around an opinion piece written by a female student studying biotechnology, Nyasha Mudukuti – one of six young scientists selected to attend the Open Forum on Agricultural Biotechnology. Mudukuti's op-ed piece, published in the *Wall Street Journal* in March 2016, argues passionately against the government's ban, which she claims amounts to "a humanitarian outrage – a manmade disaster built on top of a natural disaster."[74] Increasing pressure from both inside and outside the country seems to have had an effect: the government has already committed to importing 140,000 tonnes of maize from South Africa, which will almost certainly contain some Bt maize.[75]

Though not as visible as either Zambia or Zimbabwe, Tanzania is another country that has taken a more reticent approach towards GM crops. Tanzania's approach to new biotechnology innovation was much more cautious than its neighbours Uganda and Kenya: Tanzania was the last of the three to sign on to the Cartagena Protocol on Biosafety. Its decision makers were more influenced by the precautionary approach embodied by the African Model Law than its East African neighbours.[76] This view was particularly strong within Tanzania's Ministry of Environment, the government organ tasked with regulating new biotechnological innovations by the Environmental Management Act of 2004. This legislation cemented the country's aversion to GM crops by including a clause that put responsibility for any potential negative health or environmental consequences at the foot of the technology developer, echoing language used in the original version of the Model Law: "Without prejudice to any law governing biosafety and biotechnology, any person who develops, handles, uses, imports or exports genetically modified organisms (GMO) and, or their product, shall be under general obligation to ensure that such organisms do not harm, cause injury or loss to the environment and human health including socio-economic, cultural and ethical concerns."[77]

This resistance was entrenched within the 2009 biosafety regulations, which were the first on the continent to include both an explicit commitment to a precautionary approach, as well as a specific liability clause that attached strict penalties for any negative health or environmental outcome that stemmed from the release of GM crops, which were to be borne exclusively by the individual or organization who was responsible for the release: "All approvals for introduction of GMO or their products shall be subject to a condition that the applicant is strictly liable for any damage caused to any person or entity."[78]

This heavy-handed legislation created a chill among biotechnology proponents and scientists in the country. One account in the *Washington Post* suggested that both scientists and industry were fearful of incurring these repercussions, suggesting that no researcher dared to initiate trials of genetic modification as a result of this liability component.[79] Other accounts suggest that many officials within the government – including then president Jakaya Kikwete – worked hard to advance the case for GM crops, but that industry and scientists remained firm that moving forward was impossible with the strict liability clause in place.[80] Biotech boosters were adamant in their criticism of this bill, describing it as "prohibitive and preventative" with "strict liability provisions that are hindering technological progress."[81]

Tanzania's experimental program stalled in the years following the introduction of the biosafety law. During this period biotech boosters were working hard behind the scenes to tempt the government to overturn this policy they considered prohibitive: they convened meetings with high-level politicians to plead their case and invited high-profile speakers (including the head of PBS in 2011 and the author of *Freedom to Innovate*, Harvard professor Calestous Juma, in 2013) who rallied opposition to the bill.[82] In early 2013 the seventh Open Forum for Agricultural Biotechnology took place in Dar es Salaam, featuring a keynote address by the one of the country's most prominent biotechnologists entitled "Agricultural Biotechnology for Development," in which the speaker called for a review of the country's biosafety framework.[83] This strategy shifted in 2014 to emphasize the "seeing is believing" tours organized by biotech boosters, including two visits to a domestic biotech research facility, one tour to South Africa, one tour to neighbouring Uganda, and one to India. In June 2015, the ISAAA chose Dar es Salaam as the host of the launch of its annual report on the global status of GM crops. The audience was informed that in areas where biotechnology had been introduced, "there has been alleviation of poverty and hunger through boosting income of risk-averse small, resource-poor farmers around the world."[84] During the event, Tanzania's assistant minister for agriculture, food security and cooperatives, Godfrey Zambi, said that Tanzania "cannot afford to ignore the benefits of biotechnology in developing various sectors of the economy, especially in agriculture," and outlined the benefits of this technology, and the need for the government to support this process.[85]

These efforts finally succeeded in 2015, when the government of Tanzania removed the strict liability clause from the biosafety regulation and replaced it with a fault-based liability clause, similar to those contained in the Ugandan and Ghanaian bills. According to Hussein Mansoor, the assistant director of crop research at the Ministry of Agriculture, Food Security and Co-operatives, "this means that anyone claiming compensation for damage would have to prove that whoever introduced the GMOs was at fault."[86] But this victory was short-lived. In November 2018 the government instituted a ban on all experiments, chastising researchers for violating government procedures and ordering the immediate destruction of all ongoing trials.[87]

The skeptical approach espoused by Zambia, Zimbabwe, and Tanzania presents a major hurdle for GM proponents. Although in each country there have been internal efforts at softening its position, the pressure over the past ten years has increasingly come from the super-national regulatory level, designed to force these outliers to synchronize their regulations with the more permissive regimes of their neighbours. Zambia's decision to ban American food aid exposed how one country's biotechnology policy could significantly impact its neighbours. This controversy led to a formal request by the ministers of agriculture within the Common Market for Eastern and Southern Africa (COMESA) region for a comprehensive study of the potential impact of GM technology for regional issues of trade, food security, and emergency food aid. This COMESA-commissioned study was undertaken by the Association for Strengthening Agricultural Research in Eastern and Central Africa (ASARECA), an amalgam of the national agricultural research institutes from ten different nations.

The major recommendation that emerged from the ASARECA study called for a coordinating office to support and guide member states in their dealings with genetic modification. This program, known as the Regional Approach to Biotechnology and Biosafety Policy in Eastern and Southern Africa (RABESA), was established in 2004 to formalize regional guidelines and provide oversight on issues related to genetic modification. RABESA's financial support comes almost exclusively from USAID, with additional logistical support provided by PBS. RABESA's supporters argue that these resisters have the potential to undermine the broader trade in agriculture that makes up more than 30 per cent of the total GDP of COMESA nations.[88] Multilateral harmonization initiatives such as RABESA can thus serve to circumvent the complications that hinder the rollout of more sympathetic regulatory regimes at the national level. Ostensibly designed to rectify the unevenness and variability among national biosafety frameworks, these regional initiatives put pressure on those countries – such as Zambia, Zimbabwe, and Tanzania – that are the most resistant to GM technology to fall in line with those who have implemented more permissive regulatory regimes.[89]

THE RENEGADES

The fourth and final category of regulatory approaches is composed of those that defy categorization; those countries whose trajectories with GM crops have been shaped more by domestic idiosyncrasies than outside templates. These renegades have travelled unique paths in their engagement with GM crops.

The first such case is Kenya. Kenya began as one of the continent's most strident supporters of GM crops. One of the original participants in the UNEP-GEF process, Kenya's biosafety bill moved through parliament faster than in any of its East African neighbours; the bill was first presented to parliament in 2008 and passed into law in 2009. This was due largely to its championing by then minister of agriculture William Ruto, who was a vocal and ardent supporter of the bill.[90] NGOs and civil society organizations (CSOs) banded together in opposition, presenting an alternative draft that was introduced to parliament as a private member's bill. Minister Ruto deftly orchestrated a merger of the two bills, though the resultant Biosafety Act of 2009 represented a largely permissive framework that abandoned many of the alternate bill's key components, including provisions on labelling, strict liability, and minimum time periods for expedited review.[91] Most experts predicted that Kenya was poised to become the third country on the continent to commercialize GM crops (after South Africa and Burkina Faso).

But Kenya's auspicious trajectory was unexpectedly derailed late in 2011. The story goes like this: following the post-election violence of 2010, a power-sharing cabinet was formed that included two ministers of health with slightly different portfolios. Both of these ministers fell ill with cancer in 2011 – both sought treatment abroad and both were able to recover. During her convalescence, one of the ministers came across the now-infamous Séralini study, which suggested a link between the consumption of Bt maize and cancerous tumour growth in rats.[92] She presented this study during a cabinet meeting in November 2011 – her cabinet colleagues were similarly concerned, and, by a simple majority vote, initiated a ban on the importation and potential commercialization of GM crops until the full range of potential health implications could be studied properly. The minister of public health justified the ban on the following grounds: "My ministry wishes to clarify the decision was based on genuine concerns that adequate research had not been done on GMOs and scientific evidence provided to prove the safety of these foods."[93]

Officials within the country scrambled to reverse the cabinet decision. Biotech proponents denounced the ban as illegal, arguing that as the competent authority responsible for biotechnology, only the NBA had the ability to initiate such a drastic change in policy.[94] They convened public meetings, they lobbied relevant government officials, they advanced their position via radio, television, and op-ed pieces in local newspapers. They enlisted others to make the case on their behalf. One high-profile example was the governor of Kisumu County, who argued that the ban was undermining food security and the local economy: "it will not only harm our food security efforts but will also derail industrialisation initiatives and competitiveness of the country."[95] In another instance ISAAA and the African Agricultural Technology Foundation (AATF) recruited more than fifty smallholder farmers and brought them into Nairobi to demonstrate their opposition to the GMO ban.[96]

More recent signs point to the ban's likely demise. Having been invited to deliver the opening address at a national biotechnology conference in August 2015, now-Deputy President William Ruto dismissed the ban, suggesting that his government would have it lifted within a matter of months. Since then there has been a pivot in tactics: recognizing that the ban refers only to the importation of food and food-related products, progress in the areas of experimentation and even environmental release have been able to proceed without delays. But here again proponents were stifled by internal government resistance, as the National Environment Management Authority stepped in to block an application for on-farm testing of GM maize in October 2016, even after the country's NBA had granted its approval.[97] Although the cabinet ban still remains in place today, Kenya seems to have recaptured some of its prior momentum and now seems on the verge of becoming the next African country to commercialize GM crops.

Sudan is another country whose trajectory with GM crops defies categorization. Sudan is a bit of a wildcard in this regard; its 2012 declaration that it had become only the fourth African country to commercially release GM crops came as a shock to most observers. In the year leading up to full release there had been little investment in either scientific or regulatory capacity, largely due to the absence of formal ties with the two agencies responsible for the bulk of the continent's investment in infrastructure and capacity building, USAID and the BMGF. Sudan does have a functioning regulatory framework: the National Biosafety Act was passed in 2010 and provides a legislatively approved regulatory framework. But its competent authority, the National Biosafety Council, was established after, rather than before, commercial release in 2012.[98]

Sudan made the decision to import two Chinese Bt cotton genotypes carrying Monsanto's original Cry1A gene in 2010. These were introduced by the China-Aid Agricultural Technology Demonstration Centre at Elfaw, one of the more than fifteen demonstration centres the Chinese have established across the continent to help facilitate technology transfer in the realm of agricultural improvement. Two years of experimental trials were sufficient to convince the government of the superiority of the Chinese Bt cotton. Commercial release was approved in 2012. Two years later over 80,000 hectares had been planted with Bt cotton.

But this rapid expansion of acreage has not been accompanied by a corresponding investment in scientific or regulatory capacity. While the number of laboratories and trained experts has increased more than threefold over the past decade, as of 2014 no domestic research remains investigating rates of potential gene flow or impacts on non-target species, nor is there any investment in creating baseline measures that could be used as a comparison to measure any of these changes. Indeed, Sudan seems to have bypassed most of the established protocols that are consistent among the various biosafety laws in effect across the continent, including insufficient testing times in both the greenhouse and the field, inadequate testing on impacts of animal consumption (animals consumed foliage but not seed), and no analysis of cotton oils.[99] The only independent verification of Sudanese planting was a November 2014 visit coordinated by the industry-funded ISAAA, whose representatives reported that Sudan's path to commercialization should be viewed as a model for other countries to follow: "the lessons learned by Sudanese farmers should be shared for the benefit of Member States in the region."[100] In this sense, Sudan remains very much an outlier among the continent's nations who have embraced genetic modification as part of their broader vision for agricultural development.[101]

CONCLUSION

The divergent regulatory approaches embraced by individual African nations underline the myriad contextual variables – political, ethical, cultural, ecological, social – that make each country's reaction to these new technological possibilities unique. Still, we can tease out a number of common threads. First, regulation is occurring largely from the outside in. This may seem surprising given the typology charted above, as three of these categories – early adopters, resisters, and renegades – represent situations where domestic dynamics drove the debate more than external ones. But the category of emerging adopters remains the largest across Africa and continues to swell, encompassing countries such as Nigeria, Malawi, Ethiopia, Mozambique, Swaziland, and Cameroon. It is this blueprint of a permissive regulatory framework being driven largely by outside donors that remains the most common regulatory approach across the continent.

Second, regulatory decisions are at once technical and political. Proponents tend to portray debates around regulations as technical discussions that must be made in isolation from broader political considerations. But as political ecologist Tim Forsyth notes, "the problem of alleging such a clear separation of science and politics is to avoid the politics in the creation of the science itself."[102] This chapter highlights how decisions about regulatory frameworks reflect broader self-interest on the part of external actors, echoing political ecology's commitment to the inextricability of science and politics. Across the continent, the dominant political considerations that are prioritized are those of the organizations who are funding the creation of these new regulatory systems. Both USAID and BMGF have directed millions of dollars towards ensuring sympathetic regulatory regimes across Africa, primarily though the domestic activities of PBS and wider harmonization efforts championed by ABNE. Complex systems are designed to achieve specific goals – in this case a system of bio-hegemony is designed to facilitate the experimentation, promotion, and eventual commercialization of GM crops, which is exactly what it does.

This heavy investment has produced significant shifts in regulation across the continent. The African Model Law was adopted in 2001, espousing the precautionary principle and providing stringent regulations around risk assessment and redress in the case of hazardous release. Only a few years later, this seminal document was usurped by the *Freedom to Innovate* report, which embraced a more lenient, co-evolutionary approach "in which the function of regulation is to promote innovation, while at the same time safeguard human health and the environment."[103] By that point, the UNEP-GEF program was up and running, a program characterized by one informant as producing "common policies that would allow for easy movement of GMOs."[104] It was at this stage that donors such as USAID and BMGF stepped in and targeted countries that were more receptive to genetic modification with the promise of infrastructure, training programs, and capacity building. In turn, these early and emerging adopters became advocates themselves, lobbying more reticent neighbours to create similarly permissive regulatory structures to ensure policy harmonization.

This cycle is being repeated all over the continent. Here is one recent example. In 2016 Rwanda's minister of agriculture and animal resources made the case for expediting her country's regulatory capacity: "We need to fast track our biosafety standards to enable us deal with GMOs once they are in the country. They [East African Community members, especially Kenya and Uganda] are ahead in setting biosafety standards, while we are still defining ours."[105] A few months later, Uganda's minister of justice lobbied that the accommodation template drafted in his country needed to be adopted: "we need the [biosafety] bill because Kenya and Rwanda are planting GMOs. We believe with the law, we can fight drought; also the government has invested over USh 20bn in research, this money should not be wasted."[106] This circular logic – Rwanda needs to move forward with GM crops to keep up with Uganda, Uganda needs to move forward with GM crops to keep up with Rwanda – remains the engine propelling lax regulations across the continent.

At their core, regulations are about tradeoffs. They reflect the choices a country makes in balancing conflicting priorities, such as the desire to ensure consumer confidence in the food supply by implementing onerous restrictions that might also serve to retard innovation and the commercialization of new goods. The prevailing choice being made across Africa is one that prioritizes efficiency over efficacy, that sees promotions and regulation as complementary processes, and that uses bully tactics to pressure reluctant countries to embrace a more permissive approach towards GM crops.

SECTION TWO

GMO 1.0: First-Generation GM Crops Targeting Insect Resistance and Herbicide Tolerance in Commodity Crops

3

GM Cotton in South Africa, Burkina Faso, Sudan, and Uganda

Of all the GM crops currently under experimental or commercial cultivation in Africa, none is as well travelled as Bt cotton. Bt gets its name from *Bacillus thuringiensis*, a naturally occurring soil bacterium. What is unique about this particular bacterium is that it produces crystal-like proteins, known as Cry proteins, which kill certain insect larvae by binding to and creating holes in their gut. Most of the insects that are susceptible to Bt fall within the genus *Lepidoptera*. Known more commonly as bollworms, these caterpillars are among the most damaging cotton pests. In the 1980s, scientists at the multinational seed company Monsanto were able to identify, isolate, and extract the specific Bt genes that encode and promote the production of these toxic proteins and insert these into high-performing varieties of cotton. The resulting Bt cotton was identical to its parent in every way except that it now produced these toxic proteins, making it impervious to damage from these caterpillar pests.

This technology was designed to succeed in the large cotton producing areas of the southern United States. Throughout the 1990s it was exported to other major cotton producing countries including China and India, and then, eventually, Africa. This chapter chronicles attempts to introduce Bt cotton across the continent, with particular emphasis on its impacts on smallholder farmers. It begins with the case of the Makhathini Flats in South Africa, which was the first case of a GM crop targeting poor, smallholder farmers in Africa. It then shifts to the example of Burkina Faso, where, at its peak, over 300,000 producers were growing Bt cotton, and Sudan, which has more than 80,000 hectares under Bt cotton. It concludes with a survey of countries that are currently undertaking experimental trials with Bt cotton, including Egypt, Kenya, Ghana, and Uganda, with the aim of assessing the potential for Bt cotton to help farmers across the continent.

SOUTH AFRICA: THE CRUCIAL PRECEDENT

In 1997 South Africa passed its Genetically Modified Organisms Act and became the first country on the continent to legislate the commercialization of GM technology. Final regulations were established in 1999. The country's emergence as the continental pioneer in GM technology was spurred largely by interest on the part of Delta and Pine Land (later bought by Monsanto), which had formally submitted a request for confined field trials (CFTs) in order to demonstrate the potential of its GM cotton.[1]

Bt cotton was first approved for release in the 1997/98 growing season. Roundup Ready cotton (herbicide tolerant – or HT) was approved in 2001/02, and the stacked version containing both traits was approved for dissemination in 2003/04. Bt cotton was widely embraced by large-scale South African growers almost immediately after its initial release. Large-scale (mostly white) farmers make up the bulk of South African cotton production, and tend to account for more than 95 per cent of the country's total output. Most of these farmers are clustered in six prime growing areas.[2] Almost all of the area under cotton is irrigated, heavily mechanized, heavily capitalized, and under monoculture production. Within a few years of its release, Bt cotton was adopted by more than 90 per cent of large-scale producers, most of whom embraced the technology because of the reduced expense of pesticide application and the benefits of increased yields, which tended to be upwards of 50 per cent.

The advantages of Bt cotton to these large-scale farmers remain largely undisputed. Bt cotton is a financially remunerative insect-control strategy under these growing conditions. The more interesting story in South Africa revolves around the remote region of the Makhathini Flats, site of the first high-profile debate around the potential for GM cotton to help poor, smallholder farmers.

THE MAKHATHINI MIRAGE

Much of the debate around the potential for GM crops to help poor farmers in South Africa revolves around the Makhathini Flats, a low-lying area comprising the floodplains on either side of the Pongola

River in the far northeast of the country, near the border with Mozambique. The region is predominantly rural and very isolated. Agriculture is extremely difficult. The intense heat that dominates during the summer months makes it too dry for all but the hardiest crops to succeed. Most producers are smallholders, with holdings of only a few hectares of farmland. Almost all grow a mix of staple crops including maize, squash, beans, and pumpkin for subsistence purposes, selling any excess at the local market. This remote, rural region is among the poorest in South Africa: 90 per cent of households in the governing district earn less than 1,600 rand (US$235) per month.[3] Most households rely on pension grants from the government to supplement their income.

There have been various plans to develop the region's agricultural potential since the beginning of the twentieth century. Most of these revolved around the potential benefits for cotton cultivation, which was heralded as the ideal crop due to the region's high temperatures, low cloud cover, and sandy soils.[4] Large-scale plans – invariably revolving around new technologies and large-scale investment – followed at regular intervals: large-scale irrigated cotton plantations in the 1920s, then in the 1930s massive investment in breeding programs designed to create varieties resistant to leaf-hopping insects that ate into yields, then in the 1960s the Pongolapoort Dam was built to provide a reliable source of water for unemployed white farmers who were to be settled on twenty hectare plots, and then in the 1980s the apartheid regime relocated five thousand black farmers to try cultivating cotton along the floodplains of the Pongola River. All of these ventures failed.

It was against this historical backdrop of failure that the transnational biotech giant Monsanto introduced its patented GM cottonseeds in 1998. The first variety to be released was Bt cotton, which was targeted to Makhathini farmers whose holdings were only a few hectares. The first few growing seasons were, according to all accounts, an enormous success. A team of British and South African researchers found that, after only two growing seasons, adoption rates among smallholder farmers had jumped from 7 per cent to over 90 per cent. These researchers attributed these sky-high adoption rates to two benefits conferred by GM cotton: increased yields equivalent to a 40 per cent increase for smallholder farmers, and reduced pesticide applications, which translated into lower costs, reduced exposure, and lower labour requirements. Financial savings were substantial: on average,

Bt cotton adopters recuperated between R400 and R700 per hectare, equivalent to between US$55 and $100.[5] This research further emphasized the corollary benefits for Makhathini farmers who embraced genetic modification. Bt cotton saved women valuable time, allowing them to care for children and the elderly, or engage in other income-generating activities. Children were freed from spraying insecticide and were now able to return to school.[6] In response to the title of their lead academic article, "Can GM-Technologies Help the Poor?" the researchers replied with a resounding "yes."

Media reports were quick to seize upon the success story of Makhathini. It was trumpeted as the template for how GM technologies could help the continent's smallholder farmers. Articles published in magazines and newspapers around the world detailed how these poor farmers were finally able to afford desperately needed medicine, pay school fees, and afford basic necessities such as electricity, generators, and mobile phones due to the increased profitability of planting GM cotton. Photos accompanying these pieces showed farmers swimming in bales of cotton or showcasing the new luxuries they were able to afford thanks to the arrival of genetic modification.

No one epitomized the early successes in Makhathini more than T.J. Buthelezi, who served as chairperson of one of the largest farmers' associations during those boon years. Buthelezi was effusive in praising Bt cotton as a means of improving cotton yields. His glowing characterizations of Makhathini's first few growing seasons made him a powerful symbol of Bt's initial success in Africa: he was flown to over a dozen European countries to offer his first-hand account of the benefits he was able to reap from Bt cotton. Buthelezi was even invited to a number of private meetings with then US trade secretary Robert Zoellick, and he figured prominently in testimony offered by a high-ranking Monsanto representative to the United States Congress House Science Committee, who relayed that Buthelezi had told him, "for the first time I'm making money. I can pay my debts."[7] In each of these prestigious forums, Buthelezi was upheld as a prototypical farmer whose experiences reflected those of Makhathini farmers more generally. His unequivocal and unwavering support for Bt cotton led critics to dub him "Bt Buthelezi."[8]

The reality facing T.J. Buthelezi was much different when I visited his farm in early 2005. Buthelezi remained strident regarding the potential for Bt cotton to resist bollworms, but also grumbled about the shifting landscape among cotton companies in the area, which he complained had broken promises to farmers about price, credit, and transport. He was discouraged about the current state of his own plot: he had paid six thousand rand to have twelve hectares plowed in preparation for planting, but late rains meant that he had to hold off. By the time of my visit, Buthelezi had decided to plant this acreage with beans instead. He told me, "my head is full. I don't know what I'm going to do. I will have to speak to my wife so that we can make a plan."[9]

These gaps between the representation and reality of T.J. Buthelezi's experience with Bt cotton extend across Makahthini farmers. Figure 3.1 exposes the scale of this chasm between rhetoric and reality. What was 3,000 smallholder farmers cultivating Bt cotton in the peak growing seasons of 1999 to 2001 has shrunk to just over 500 over the last two years. In terms of acreage, 255 hectares were under cotton in 2013/14, which represents a decline of more than 90 per cent from the peak acreage achieved in 2001/02. Total production dipped from a zenith of over 8,000 bales to around 1,000 in 2013/14.

So how do we explain the dramatic rise and precipitous drop of Bt cotton in Makhathini? The trajectories of Figure 3.1 are best explained not by the technology itself but by the changing context in which it operates. As I have argued elsewhere, new agricultural technologies and their contexts cannot be understood in isolation: their outcomes in any particular place are determined in large part by the specific institutional, economic, and political circumstances that frame their introduction. My research in Makhathini incorporated extensive conversations with farmers alongside historical analysis and institutional data, and reveals a number of institutional, economic, and environmental factors that buoyed Bt cotton after its initial release and precipitated its downfall.

Let's deal with the rise of Bt cotton first. A number of factors played significant roles in accelerating the adoption of Bt cotton in the period immediately after its commercial release. The first is geography. Makhathini is remote and isolated: it is bounded to the east by the Indian Ocean, to the west by the Lebombo Mountains, and to the north by the border with Mozambique. This leaves farmers with few options for growing crops for sale. The region's cotton gin represents the region's sole commodity market, leaving cotton as the sole option for farmers wishing to cultivate a commodity crop. Conversations with Makhathini farmers suggest that the biggest

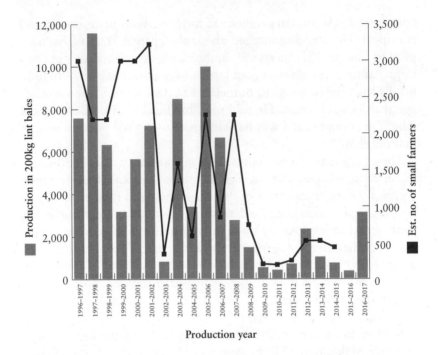

Figure 3.1 Estimates of cotton production by smallholder farmers in KwaZulu-Natal. Original figure created based on data obtained from Cotton South Africa, "Textile Statistics" (database), http://cottonsa.org.za/resources/textile-statistics.

single advantage conferred by Bt cotton is the close, stable market it provides to producers. This guaranteed market played a critical role in sustaining adoption rates; these rates cannot be explained entirely by the yield increases or pesticide reductions that accompanied the introduction of Bt.

Institutional factors also played an important role in buoying adoption rates following the initial release of the technology. From 1998 to 2001 there was a single cotton buyer in Makhathini, Vunisa Cotton, whose responsibilities included selling seeds and inputs, buying cotton, offering extension services, and providing credit in conjunction with the South African Land Bank. The company used its monopsonistic power to supply credit to farmers who did not own their land by allowing the forthcoming crop to be used as collateral – the amount owed was deducted before paying farmers for their output. Farmers suggested that many people adopted Bt cotton as means of accessing credit supplies, and that Vunisa's privileged position as the region's sole provider of credit was a key factor that helped account for the rapid takeoff of Bt cotton in 1998. But Vunisa's model collapsed in 2001, and its generous credit program along with it. Not surprisingly, adoption rates plummeted by more than 90 per cent in the subsequent growing season (see figure 3.1). By most accounts, Bt cotton's dramatic rise was buoyed by supply more than demand. Adoption rates were shaped by Vunisa's business strategies, lending capabilities, and privileged position as the sole cotton buyer in Makhathini, more than by any need or desire voiced by farmers.[10] The high adoption rates recorded during these years reflect the wide availability of credit as much as they do demand for Bt technology.

This link between adoption rates and credit availability persisted throughout the decade. The only farmers who were able to continue with cotton once the credit dried up were the elderly, who were able to finance input by using their pension income.[11] In 2005/06, after more than two years of protracted negotiation, the KwaZulu-Natal Department of Agriculture stepped in to subsidize inputs by making more than 6.4 million rand available as "bundles" designed to improve farmer access to the more expensive Bt seed. Predictably, the number of farmers planting Bt cotton spiked, from 548 in 2004/05 to 2,169 in 2005/06, only to rebound back to 853 once the program was discontinued in 2006/07.[12] The wide swings in adoption rates during this first decade of adoption reflected these fluctuations in credit availability more than they did any ecological or economic advantage of the seed itself.

Some observers have suggested that the sudden credit vacuum that opened up following the collapse of Vunisa's lending operations and the subsequent decline in cotton adopters would have occurred irrespective of Bt technology.[13] That is, the heavy fluctuations accorded to credit are a characteristic of the cotton crop, regardless of whether it is genetically modified. But Bt cotton requires this high credit context more than its non-GM cousins due primarily to the price differential between the two. In the first few years of adoption, Makhathini farmers were protected from this price differential and were only made to pay a technology fee of 230 rand for a twenty-five kilogram bag, which was less than half of that charged to larger-scale farmers. But similar to the availability of credit, these generous price differentials did not last, and within a few growing seasons Makhathini farmers were required to pay the same technology fee as other farmers across

South Africa (785 rand for Bt, 365 rand for Roundup Ready, and 1,150 rand for the stacked variety), making the price of Bt cotton more than double that of non-Bt seed. Fok et al. estimate that conventional seed represented somewhere between 40 to 60 per cent of a farmer's total input cost, while the added technology fee associated with Bt cotton increased this figure to 70 to 80 per cent. This made Bt cotton farmers more reliant on credit to cover these increased costs accrued at the start of the growing season and less able to continue once the credit dried up.[14] Another estimate suggests the gap between the price of conventional and GM seed is much higher, nearly three times as much ($2.55/kg versus $7.20/kg).[15] These arrangements create greater financial risk for the farmer as a greater share of the overall cost is incurred prior to planting and thus cannot be adjusted later in the season.

Other structural factors played a part in privileging the adoption of Bt cotton. In the wake of Vunisa's collapse, a second firm, the Makhathini Cotton Company (MCC), stepped in to fill this gap as the region's sole seed purveyor and cotton processor in 2002. The MCC sought to continue and extend Vunisa's business plan with some slight alterations: it decided to sell seed not directly, but rather through an intermediary set up on-site, and, unlike Vunisa, it decided not to offer credit directly to farmers, opting instead to facilitate access to credit via the Land Bank. The MCC implemented a series of initiatives designed to buoy Bt cotton adoption in Makhathini. First, the MCC mandated that it would only buy cotton delivered in specially marked woolsacks. These sacks were allocated only to farmers who opted to plant Bt cotton at the beginning of the growing season, eliminating the local market for any farmer who opted for non-Bt seed. The MCC constrained farmer choice by only making non-Bt cottonseed available in quantities of twenty-five kilogram bags – enough to seed ten hectares of cotton, meaning it was too expensive and too much seed for the smallholder farmers who predominate in Makhathini. Bt cottonseed, on the other hand, was marketed in an "ecombi" five kilogram package, an ideal size for the area's smallholder farmers. Still, MCC struggled to achieve viability, which hinged on successfully ginning a minimum of ten million kilograms annually. To boost production numbers even further, the MCC created joint-venture schemes in which the MCC would take control over cotton production in order to implement joint-pivot irrigation and take advantage of economies of scale.[16] The creation of these large plots of contiguous lands under cotton, which involved forcible evictions of farmers who were reluctant to join such schemes, were a long way from the original depiction of Bt cotton as a technology capable of empowering smallholder agriculture. The high adoption rates recorded in Makhathini are thus due in part to these restrictive buying and selling arrangements.

The MCC closed up shop in 2007. As figure 3.1 shows, the production of Bt cotton in Makhathini has stagnated to negligible levels since then. The Department of Agriculture and Fisheries bought the gin formerly owned by Vunisa and then by the MCC for thirty million rand in 2010. It remained closed for two years of repairs before reopening in 2012.[17] The reopening of the gin was hailed as having momentous promise: one report hyped that the gin would "transform the rural community."[18] The new plan for cotton expansion revolves around the creation of an "agri-park," a form of P3 that seeks to locate the entire cotton value chain in the area.[19]

How then do proponents explain the drop in numbers after the 2001/02 growing season, and the persistent low production values that have dominated since that time? Some researchers continue to celebrate Makhathini's success, suggesting that farmers are better off with Bt cotton than they were before. Others have reconciled the recent declines in both overall production and adoption rates, arguing that fluctuating weather conditions and poor institutional structures undermined the long-term prospects of Bt cotton.[20] These accounts stress that it is the institutional arrangements, not the technology, that are flawed, suggesting that the lesson learned from Makhathini is that "farmers can benefit from technological innovation only if the correct infrastructure is in place."[21] Within this view, the experience in Makhahini with Bt cotton is characterized as a "technological triumph but institutional failure."[22]

My analysis disputes these conclusions. As political ecology makes clear, technologies and their individual contexts cannot be understood in isolation; GM crops are only as successful as their interaction with their surroundings allows them to be. Yet the perpetuation of the narrative of Bt cotton's technological promise relies precisely on this illusionary separation of technology and context, suggesting that the technology is sound but the implementation model is flawed. As Dominic Glover argues, "the positive evaluation of Bt cotton's socio-economic impacts has depended on a reductionist analysis, which has sought to detach the technical performance of the Bt trait from the agro-ecological and institutional context where it has been put to use."[23]

The site-specific analysis undertaken above seeks to understand how ecological, institutional, and economic variables contribute to the relative success of agricultural biotechnology.[24] The omission of such factors in underpinning Makhathini's success is an essential part of the process of technological storytelling: it serves to perpetuate the myth of success, thereby extending the significance and promise of this technological intervention.[25] Detailed, empirical studies are needed to unravel the specific conditions that frame the introduction of GM crops in order to expose the gap between aspirations and experience.

When Makhathini's geographical and institutional contexts are taken into account, the dominant representation of Makhathini as a prototype for how GM crops can elevate the position of Africa's rural poor collapses. In its place a more vulnerable view of Makhathini emerges, one in which certain preconditions exist that have significantly contributed to the success Bt cotton has enjoyed there: geographical isolation that eliminated the possibility of alternative crops, and monopsony structures that encouraged production by providing easy credit, in the case of Vunisa, or excluded the potential for non-Bt cotton and profited from economies of scale, in the case of the MCC. Bt cotton is a technology that requires a certain context in order to succeed: large-scale, heavily capitalized, with credit widely available. These conditions were satisified in Makhathini for the first few years after the release of Bt cotton, creating a context in which the technology was able to thrive. But sustaining these monopsony market structures and profligate lending strategies over the long-term proved impossible. Such assessments fail to appreciate the mutual inseparability of GM technology and the context required to sustain it.

BURGEONING BURKINA

Even once the Makhathini experience had been exposed as a mirage, other countries adopted Bt cotton and emerged as the new darlings of proponents of genetic modification, offering new evidence of the potential benefits that GM cotton can offer to African farmers. As in the case of Makhathini, though, a closer examination of the circumstances that facilitated these successes demonstrates how important social, economic, and ecological factors are to these representations of success. More crucially, exposing the full picture of events raises important questions about the ability of these technologies to help poor farmers over the long-term.

Landlocked in the subtropical zone of western Africa, Burkina Faso has languished as one of the world's poorest countries since its independence in 1960. The population remains predominantly rural, with more than 85 per cent of citizens engaged in agriculture. Cotton dominates as the country's most important commodity crop. Cotton is of enormous importance to the economy as a whole: it accounts for more than 10 per cent of the nation's GDP, generating revenues in excess of $300 million (representing more than half of the country's export earnings).[26] The vast majority of the nation's 350,000 cotton producers are small scale: over 90 per cent of farmers have total holdings of under five hectares, with average cotton production ranging between one and two-and-a-half hectares. Yet despite these small acreages, cotton remains central to producer livelihoods, accounting for more than 60 per cent of average incomes.[27]

Bt cotton was first introduced in Burkina Faso in 2008. As in South Africa, preliminary adoption rates were sky-high: more than 100,000 smallholders were growing Bt cotton within a few growing seasons. Acreage expanded steadily from an initial planting of 9,000 hectares to just under 500,000 hectares in 2013.[28] Soon after, Burkina Faso was declared as a "role model" that other African countries could emulate in using GM technology to benefit smallholder growers.[29] Burkina Faso supplanted South Africa as the most important case demonstrating the potential for GM crops to help poor farmers.

In order to understand the rapid rise of Bt cotton in Burkina Faso, one must first grasp the history of cotton as the country's primary commodity crop. This story of cotton production in French West Africa is one of increasing penetration, consolidation, and regulation. Although cotton production was firmly entrenched in West Africa before the arrival of French colonialists, it was during the colonial period that cotton production transitioned from primarily serving domestic tastes and regional markets to a more standardized mode of production designed to satisfy increasing demand from European industrialists.[30] During the colonial period Burkina Faso's cotton production became more centralized and more dependent on state institutions for its operations. Independence from France in 1960 ushered in a boom period that saw surging levels of production through the 1960s and 1970s, fuelled in ways both extensive (via the government's support for animal traction and the increasing penetration of cotton into the country's southwest, which resulted in a tripling of total land under cultivation) and intensive (via investment in inputs, extensive services, and yield-enhancing research, which resulted in yields quadrupling over these decades).[31] The late twentieth century was also a period of considerable improvement in the quality of cotton produced, due primarily to the work undertaken by a series of French research institutes, whose main goal was to create cultivars that were well adapted to the growing conditions in West Africa.

By the 1980s Burkina Faso's cotton was renown for both the quantity and quality of cotton produced; it was considered to be the second best in on the continent and to be of better quality than most machine-picked cotton in the United States.[32] But by the 1990s the industry was threatened by increased resistance among insect pests – among them the larvae *of H. armigera* (a bollworm from the genus *Lepidoptera*) – to the routine spraying of pyrethroids and organophosphates by farmers, causing farmers to increase their pesticide load by more than 25 per cent, which in turn increased costs and exposure.[33] Bollworm damage was reported to be as high as 40 per cent even when a full application of insecticides was used.[34] Initial management efforts revolved around the window spray approach, which involved coordinated times for regular sprays and enhanced farmer scouting. But these agronomic management techniques did little to stem declining yields and spiralling production costs. Burkinabè officials consulted experts from around the globe to help them manage this crisis of mounting resistance. In July 2003, the government decided that genetic modification was the most promising means of overcoming the obstacle of insect resistance, and invited companies holding licences over GM cotton to demonstrate their potential in Burkina Faso.

The rapid success of Bt cotton in Burkina Faso was bolstered by the country's tightly regulated and highly integrated system of cotton production. All cotton growers in the country are organized into collectives of fifteen members, known as GPCs (Groupement des producteurs de coton). This collective representation then scales up into local, provincial, and national unions, known as the UNPCB (Union nationale des producteurs de coton du Burkina). According to the head of the UNPCB, this collectivist model facilitates communication, harmonizes practices, and offers access to cheaper inputs and makes credit more available (in the 2014 growing season, UNPCB was able to access over CFA 3 billion in loans).[35] This collectivist structure also facilitates representation: instead of dealing with each cotton grower individually, cotton agents deal exclusively with GPCS and their higher-level representatives. This ensures a higher level of accountability than a free market could offer and allows the cotton companies to provide generous advances based on the collective's guarantee.

In addition to UNPCB and the conglomerate of collectives that it represents, two other national organizations are at the core of this centralized apparatus. The first is INERA (Institut de l'environnement et recherches agricoles), Burkina Faso's national agricultural research centre, tasked with overseeing the country's experimental program into improved cotton varieties. The second is SOFITEX (Société burkinabè des fibres textiles), the largest and most powerful of three parastatal cotton companies whose responsibilities cover almost every link in the cotton value chain, including providing inputs, buying cotton, undertaking ginning and commercialization, and financing research.[36] Together, these three organizations constitute the AICB (Interprofessional Cotton Association of Burkina), an umbrella organization that represents all major stakeholders within the national cotton industry.

These three layers of institutions operate together under what is known as the Integrated Sector Approach (l'approche filière intégrée). At the start of every growing season, cotton companies such as SOFITEX provide all the inputs to each GPC, including seed, fertilizers, herbicides, and pesticides. Once the harvest is complete, the GPC sells the harvested cotton back to the companies, who 1) deduct the various expenses incurred at the start of the season and 2) pay a flat fee of CFA 5 per kilogram directly to the UNPCB. By dealing exclusively with the GPC, the cotton companies avoid any of the low repayment rates that plagued operations in South Africa: if one farmer's crop underperforms, making him unable to pay back the costs associated with his inputs, this amount is covered by the profits made by another member of the GPC. This "caution solidaire" (collateral security) helps to mitigate loan defaults by spreading the responsibility for payment across the collective.

It was the AICB who first reached out to Monsanto and invited their representatives to conduct a series of workshops with government officials and farmer leaders in Burkina Faso in 2000 and 2001, introducing them to the potential advantages of Bt cotton as a means of overcoming the high levels of pesticide resistance that had become endemic by the late 1990s. Monsanto officials made it clear that

Burkina Faso needed a strong regulatory structure in place before they could release their proprietary technology:

> when we first started discussions and essentially told them we can't really pursue anything in Burkina Faso unless we feel there is a regulatory environment that would allow us to do so, and the powers that be, the key players in some of the ministries worked very hard to develop what we would consider some biosafety legislation that would allow us to work in the country. It wasn't restrictive. And they worked very hard to quickly set up competent authorities to regulate the testing and commercialization of biotech products.[37]

In 2001 SOFITEX initiated a partnership with Monsanto to explore the potential of using its Bollgard II cotton containing both Cry1Ac and Cry2Ab genes as a means of overcoming stagnating yields and ever-increasing pest control measures. This collaboration was formalized in 2003, inaugurating five years of field testing on experimental stations and select farms, supervised by INERA scientists.[38] A biosafety committee comprising experts from the various implicated sectors was established in 2003 to oversee these experiments. In 2006, the country passed its biosafety law establishing the National Biosafety Agency as the focal point for regulating and safeguarding both the experimentation and eventual commercialization of GM crops.

Initial enthusiasm was propelled by results obtained during three years of CFTS (2003–05), during which two of Monsanto's patented Bt genes in two American germplasms (DP 50 and Coker 312) were tested in two separate CFTS. This research – steered by INERA in partnership with the technology provider, Monsanto – found that the Bt cotton produced significantly higher numbers of bolls. Yields increased by 15 per cent and insecticide applications were cut by over two-thirds. Economic modelling suggested that Bt cotton would offer economic gains of between US$35 to $110 per hectare (depending on the price of seed).[39]

Although the results of these initial tests were positive, Burkinabè officials expressed reluctance over the importation of American germplasm, which they considered to be inferior to their own domestic varieties. The Burkinabè cotton industry was fiercely proud of the reputation of the cotton it produced, which was generally considered to be the best on the continent after Egyptian Pima. Burkinabè officials expressed these concerns to their collaborators at Monsanto, insisting that the Bt event needed to be inserted into their local varieties, which were high yielders and well adapted to local growing conditions, and that no commercial release could proceed until this occured. In the words of one Monsanto official: "the government officials were not interested in us bringing our varieties in. So we decided to work with them and introgress the traits that crossed the traits into their local germplasm because number one they wanted it that way and it preserves the agronomic qualities that they valued locally."[40] Monsanto consented, and in 2006 the Bt genes were backcrossed and transferred into the country's three most popular varieties: FK 37 and FK 290, which dominate throughout the country, and STAM 59 A, a Togolese variety that is grown in areas with lower precipitation levels.[41] That same year, these newly generated GM versions of local germplasm were pre-released among twenty large-scale farmers, who reported an average yield increase of 20 per cent in the first growing season. After three generations of backcrossing, the company announced that the new GM lines were stable and ready for commercial release.[42] This was sufficient for the country's National Biosafety Agency, which granted approval for nationwide commercial release in 2008. Cotton distributors planted 15,000 hectares in order to generate sufficient seed supply. In 2009 over 125,000 hectares of Bt cotton was planted in the country, the most extensive single-year planting of any GM crop on the continent to date.[43]

The data from these first few years of mass planting were extremely promising. The first study undertaken included 160 rural households spread across ten different villages. They found that Bt cotton increased yields by an average of 22 per cent. Higher seed costs (Bt seed cost four times as much as conventional seed) were offset by reduced insecticide costs (these decreased by almost one-fifth). This translated into sizeable economic benefits: an average of $65 per hectare over conventional cotton, meaning that a farmer with average holdings saw an increase in household income of over 50 per cent.[44] Put another way, the production costs associated with a hectare of Bt cotton was 14 per cent lower than a hectare of conventional cotton over these three growing seasons. Aggregated over a national level, total additional revenues generated by Bt over the course of these three years was calculated to exceed US$50 million dollars.[45]

What is more, this initial research suggested that these net gains were accruing disproportionately to the poorest and most vulnerable farmers. Researchers found that smallholders without access to animal traction or other labour-saving technologies benefited more from the introduction of Bt relative to medium- and large-scale producers (27.9 per cent, compared to 17.7 per cent for large producers and 7.7 per cent for medium-size producers).[46] This led the authors to conclude that Bt cotton had played an important role in evening the playing field: "Bollgard II has enabled manual producers, by adopting Bt cotton, to compete with the larger, better equipped, and often more skilled conventional cotton growers."[47] This research also showed that producers were able to secure over 46 per cent of these benefits, the rest going to cotton companies and the technology provider (Monsanto).

My own experiences with cotton growers in Burkina Faso corroborate these reports of success. I visited Bereba District, in the western part of the country near the city of Bobo-Dioulasso, at the start of the planting season in 2015. Bereba, in the heart of the Burkina Faso's cotton-growing district, is a dry, brown landscape, dotted by large trees. Soils are clayish and dry. At the time of my visit, most farmers had planted a third of their plot and were waiting on the rains in order to put the rest of their lands under cotton. Cotton that had been planted one or two weeks before had still not sprouted, and those that had were wilted and dried out by the sun. Despite these difficult conditions, farmers spoke very positively about Bt cotton. When I asked whether farmers preferred Bt cotton over their conventional varieties, one farmer smiled and replied that Bt cotton is so good that, even if a farmer had difficulty facilitating transport to the local gin, buyers would come from the capital in order to buy it.

All the farmers I spoke with confirmed that their yields were higher with Bt cotton: the consensus was that Bt cotton outyielded conventional cotton by approximately 10 per cent. Specifically, farmers noted that a typical Bt cotton plant could produce as many as fifty to sixty bolls, while a conventional plant tended to max out at twelve or fourteen. Farmers also confirmed the claim of reduced pesticide applications: most went from six sprays per season down to two, which represented an important savings in terms of money (each spray costs CFA 4,000), time (each spray requires a farmer to walk approximately 15 km back and forth for every square kilometre planted), and health (farmers reported less pesticide exposure).

The above narrative is the story of Burkina Faso's success that has been widely trumpeted as proof that Bt cotton can help poor farmers across Africa. But cracks have begun to emerge in this finely painted portrait. Bt cotton accounted for 70 per cent of all cotton planted in 2014, which represents the upper threshold in terms of coverage (30 per cent needs to be devoted to non-Bt cotton varieties that act as refuges to stave off the development of pest resistance to the Bt construct). But this ceiling was reached only once. In 2016 the proportion of land under Bt was reduced to 50 per cent, then reduced again to 30 per cent in 2017. By 2018 the country had undertaken a complete return to non-GM seed. As one of the country's leading cotton experts put it to me when we met in Ouagadougou: "[Bt cotton] is no longer a success story in Burkina."[48]

What is behind this systematic reduction of Bt cotton? The answers revolve around the precarious issue of cotton quality, one of the most important traits distinguishing West African cotton from its international competitors. The term "quality" refers to two groups of related phenomenon. The first relates to the features of the fibre. Burkinabè cultivars are the product of decades of careful breeding that has resulted in premium cotton fibres that are long, strong, and uniform. These traits are highly sought after for the production of high-end textiles and fetch a premium on the global market. The second reason why Burkinabè cotton fibre is of such high quality stems from it being handpicked, which ensures that the fibre is free of other organic matter.[49] Handpicking influences the other quality trait valued by cotton companies, known as the ginning ratio, which is the percentage of fibre per unit weight of cotton delivered to the gin. The ginning ratio of Burkinabè cotton is high, the result of decades of targeted breeding and careful handpicking. A high ginning ratio is attractive to Burkinabè cotton companies since it increases the total amount of fibre that it can sell at a high value compared to the total harvest weight. Burkinabè cotton has gained a stellar international reputation and a premium price based on these quality traits.

The high quality of Burkinabè conventional cotton is the result of a very successful breeding program that was initiated by the French colonialists.[50] In 1946, the French government founded the Institute for Research on Cotton and Tropical Textiles, known by its French acronym ICRT, to lead cotton-breeding programs for its African colonies. The breeding efforts began by ICRT became part of the French agricultural research organization, CIRAD, in 1984, and were eventually absorbed into national research institutes. The main goal of the ICRT-CIRAD breeding program was to create cultivars that were well adapted to the growing conditions in West Africa, exhibiting desired quality characteristics such as a high ginning ratio and long staple length.[51] This breeding program achieved considerable success. Since the 1970s, the average ginning ratio for Burkinabè cotton increased from 36 per cent to 42 per cent in 2006/07. The improved ginning ratios in the Francophone West African cotton sector in general, and the Burkinabè cotton sector in particular, were to a large extent the distinguishing feature that made them more competitive in the global market. The ginning ratios in other African countries were unable to match this progress over the same period of time. The cotton-breeding program in Burkina Faso also made considerable improvement in staple length over this time period. The standard benchmark in the cotton industry for the more desirable medium-to-long cotton fibres is one and an eighth inches. The percentage of total Burkinabè cotton classified as exceeding this threshold rose from 20 per cent of total cotton production in 1995/96 to 80 per cent in 2005/06.[52]

Both Monsanto breeders and Burkinabè stakeholders knew there was a chance that introgressing the Bt event into the local Burkinabè varieties could interfere with these two highly valued quality characteristics. But the three years of CFTs revealed no such interference and "the fibre's characteristics were maintained."[53] In the words of one Monsanto official familiar with the trials, "All I can say is, based upon the assessments that we made with their help, we were achieving things that were satisfactory to them at the time."[54]

Concerns over quality emerged in the first year of commercial release. Breeders noticed that the GM versions of the FK 26 and FK 90 varieties were producing ginning ratios inferior of 42 per cent, as well as a lower fibre length value.[55] Monsanto officials were skeptical, suggesting that these initial declines in ginning ratios and staple length were due to exceptional water stress and other climatological variations.[56] But this deterioration in ginning ratios and staple length persisted over time. Reports from Burkinabè officials, which were corroborated by Monsanto, confirm that Bt cultivars produced fibres that were one-thirty-second of an inch shorter than conventional varieties.[57] In the 2013/14 season, over two-thirds of the nation's total crop was classified as lower-to-medium quality (with a staple length between one and three-thirty-second inches and one and one-sixteenths), with only a third retaining its previous classification as medium-to-high staple length. This represented a decline of over 40 per cent since 2005/06.[58] The precise decline in ginning ratios is more difficult to measure, though Burkinabè officials confirm that it remains well below the 42 per cent achieved by conventional cultivars.[59]

This decline in staple length has undermined the reputation of Burkinabè cotton and cut into its value on the international market. When coupled with the decline in overall lint due to the lower ginning ratio, the inferior quality characteristics of the Bt cultivars have compromised the economic position of Burkinabè cotton companies. The lower-quality fibre was valued less highly by spinners, who could only use the poorer grade for the production of lower-quality textiles, such as bedding. It also complicated trading arrangements among other West African producers such as Côte d'Ivoire and Mali. All Francophone West African producers aim for a homogenous product that can be interchanged to facilitate timely delivery to clients, but Burkina Faso's poor staple length undercut this flexible sourcing mechanism.[60] In 2015, Burkina Faso produced over 700,000 MT of cotton, while its western neighbour Mali produced only 500,000 MT. Yet within a few months Mali's entire product had been sold on the international market, while most of Burkina Faso's languished, awaiting export. As one high-ranking official lamented, "What is the point in being the top producer if you can't even sell your cotton?"[61]

Breeders are struggling to account for these declines in ginning ratio and staple length. In theory, inserting the Bt gene into the Burkinabè germplasm should have left the resultant progeny identical to its parent in every way except for the inserted trait conferring insect resistance. But in reality, the process of introgressing the Bt trait into the local variety appears to have interfered with some of its most important characteristics. Monsanto scientists are at a loss to explain the precise mechanism that has created these problems: "I mean, we don't know exactly what is going on – but for sure you have genetics and environment, blaming some factor coming from genetics and environment – which are reducing the length of the fibre."[62] The company is attempting to identify and correct this fault. In the short term, Monsanto has proposed forming a technical committee of local and international experts to investigate this issue of declining quality and come up with recommendations for moving forward. In the medium- to long-term, Monsanto has embarked upon a new process of backcrossing the Bt trait into a new local cultivar, known as FK64, which, according to Monsanto's internal trials, seems immune to these deteriorations in quality. The company promises to pay more attention to fibre quality when undertaking these backcrosses and to use "new tools and processes" to ensure that the resulting backcrosses do not suffer similar deteriorations in quality.[63]

Burkina Faso's cotton companies have grown impatient and have decided to take matters into their own hands. Frustrated with Monsanto's inability to identify and correct these declines in quality, the companies set a timeline for abandoning Bt cotton and returning to conventional Burkinabè cultivars. Their centralized control over the country's seed supply allowed them to reduce the availability of Bt cottonseed from the peak rate of adoption of 73 per cent in 2014/15, 53 per cent in 2015/16, 30 per cent in 2016/17, and a complete return to conventional cotton for the 2017/18 season. The cotton companies also made a formal request to Monsanto for losses incurred due to these declines in quality, demanding more than CFA 48 billion (approximately US$76 million) as compensation for losses sustained since 2010. One high-ranking Burkinabè official articulated the industry's position as follows: "As a result of their negligence, we have no other option but to take care of our own responsibilities while we wait to be compensated; we will phase out Monsanto's genetically modified cotton. We will gradually reduce the amount of land allocated to their gene … This is all a result of Monsanto's inability to find a solution."[64]

This dispute was resolved in early 2017, with Burkina Faso's cotton sector and Monsanto agreeing to a 75/25 split of all remaining royalties. SOFITEX's managing director stated that this represents the termination of their business relationship: "In doing this, we think that a bad deal is better than a bad court case. We have closed the Monsanto dossier."[65]

What lessons can we draw from Burkina Faso's volatile experiences with Bt cotton? The first takeaway is that Burkina Faso's early success was due in large part to a favourable context for adoption. The most important work in this area has been undertaken by Brian Dowd-Uribe, whose doctoral dissertation investigates the years leading up to and following Bt cotton's widespread release. Dowd-Uribe argues that Burkina Faso's vertically aligned, state-controlled monopoly was attractive to Monsanto due to their previous frustrations with unruly seed markets in India and the resultant complications around pirated and fake seeds.[66] This venture was viewed with equal enthusiasm by INERA, who were going through a period of severe government cuts – chronic underfunding that had led to persistent delays in salaries.[67] The agreement with Monsanto set out that 28 per cent of the royalties from the sale of seeds would go to them, while the rest would be split among the cotton companies and INERA. The potential royalties include both domestic and export production; Burkinabè cotton officials were particularly excited about the possibility for exports to neighbouring West African countries including Ghana, Côte d'Ivoire, and Togo. It is not surprising, then, that cotton officials in Burkina Faso were keen to partner with Monsanto to examine the potential for GM cotton to improve the nation's cotton production.

My own conversations with Burkinabè officials corroborate this point. Monsanto officials were upfront about the benefits of the pre-existing vertically integrated system, which made it "easier to capture for our investment" and lowered the potential "leakage" of technology via farmer-to-farmer channels.[68] The parastatal structure allowed Monsanto to overcome one of the major stumbling blocks that had stymied the dissemination of Bt cotton among high numbers of small-holder growers in India: the unwieldy problem of selling, contracting, and enforcing legal compliance to restrict reselling and reusing Bt cottonseed among a large population of farmers. The vertically controlled industry allowed Monsanto to avoid dealing directly with producers, signing instead only a single contract with the Burkinabè cotton industry.[69]

The second lesson concerns the narrow scope of the insect-resistance breeding program. Bt cotton was originally bred in the United States with the sole aim of conferring the Bt trait into a cultivar that would express the toxin consistently. This exclusive focus on pest mitigation contrasts sharply with the Francophone West African breeding programs, which spent decades integrating a broad spectrum of adaptability to growing conditions alongside multiple characteristics of fibre quality. The Burkinabè cotton industry astutely tried to remedy the undesirable characteristics of the American cultivar by backcrossing it into its own cultivars. Three generations of backcrossings were undertaken, which is standard practice in the United States where quality issues are much less pronounced due to the heavy reliance on mechanized pickers. But in Burkina Faso, where quality concerns are paramount, some breeders advise a minimum of five generations of backcrossing to ensure the carryover of the desired beneficial traits.[70] As a result, the desire for stability and quality clashed with the desire to get to market, as each backcrossed generation takes a year of careful breeding and selection. As a result, quality suffered. The process of introgression is complex and time-consuming, and potential conflicts can emerge when the priorities of private patent holders clash with those of other actors. This stunted breeding program calls into question the potential for combining GM technology and local cotton cultivars to produce new technologies that offer desired performance across multiple criteria, as well as focusing on the GM trait rather than the suite of characteristics of the germplasm into which it is conferred.

A third set of questions revolves around the nature of Burkina Faso's pullback from Bt cotton. In Burkina Faso, the decision to phase out Bt cotton was made by the cotton companies rather than cotton farmers. Burkinabè cotton companies were frustrated by the declining profits associated with the poorer lint quality of Bt cultivars. Farmers had no choice in this matter: thus, the farmers with whom I spoke in Bereba were keen to continue using Bt cotton and anxious about the decision to discontinue the use of Bt cottonseed – however, due to SOFITEX's total control over the seed distribution and buying system they have no choice in the matter. Both farmers and cotton companies benefit from a vibrant and profitable cotton sector; the cotton price paid to farmers is ultimately a function of the price at which the cotton company sells it on the world market. In this particular case, though, the interests of the companies and the farmers diverged: the higher yield of Bt cotton meant more income for farmers while the lower ginning ratio and shorter staple length meant less fibre and a lower-quality cotton for cotton companies to sell. What is more, cotton officials were not willing to divulge the true reasons for the pullback to farmers, arguing that withholding this information was for their own good:

> We need to reduce the area delegated to Bt seeds, but we can't just tell producers that we are reducing the quantity of Bt seeds used because they won't understand. Instead, we have to tell them that there are no more Bt seeds. We have to say that there was an issue, and as a result we have no more Bt seeds. Instead, we suggest that the producers use conventional seeds. This is because we have to try to restart. This is the message that we are passing along because it's too difficult to explain the reality to producers. Their superiors are up to date and they will progressively inform the producers. Why? Because it is a process.[71]

The case of Bt cotton in Burkina Faso exposes the conflicting interests within the cotton value chain, underlining how GM crops can produce different outcomes for different stakeholders. As in Makhathini, a close examination of the institutional landscape that framed the dissemination of GM cotton reveals the fundamental relationship between technology and context. In this sense, then, Burkina Faso is a powerful counterpoint to the Makhathini example, which supporters characterized as a technological success but institutional failure. In Burkina Faso, Bt cotton was an institutional success but a technological failure.

THE SITUATION IN SUDAN

In addition to South Africa and Burkina Faso, Sudan is the only other country currently growing GM cotton on the continent. Sudan is unique among African countries in that it has a long history of large-scale cotton cultivation. These efforts date back to the British Gezira Scheme, a massive plateau of uniform fields growing irrigated cotton in the alluvial basin just south of the confluence of the Blue and White Nile. This was the largest colonial cotton project ever embarked upon, involving the damming of the Blue Nile, the creation of a vast system of canals carrying water to many thousands of hectares, and the forcible enlisting of tens of thousands of tenants whose farming activities were dictated by the central administration. At its peak, the scheme administered over 87,000 hectares covered with cotton, with more than 25,000 households compelled to cultivate it.[72]

The colonial chapter of cotton production concluded with independence in 1956, at which point the Sudanese state inherited and expanded this vision for large-scale cotton cultivation. Cotton was an important lynchpin of the nascent Sudanese economy, accounting for more than 65 per cent of foreign exchange earnings at its peak. A cocktail of different insect pests – including bollworm, jassids, aphids, and white moths – emerged as the major hindrance to production efforts. The most cost-effective mechanism for controlling these pests was aerial spraying of organophosphates. These were applied en masse, peaking at twelve sprays per season. The effectiveness of these pesticides began to decline in the 1970s as resistance emerged, forcing farmers to switch over to pyrethroids. While these were effective in killing targeted pests they precipitated a near collapse of the entire industry, as they killed off all the natural predators of the white fly, which, left unabated, wiped out nearly the entire crop in the early 1980s.[73] By the 1990s production had rebounded, with more than 400,000 hectares under cotton across the country. Pest control has persisted as the single most important obstacle to cotton growing, accounting for between 30 to 40 per cent of all production costs.[74]

Bt cotton is the latest in this long series of pest control efforts. Two years of experimental trials comparing two Bt varieties imported from China against local germplasms revealed the "consistent superiority" of the Bt varieties. Bollworm damage was greatly reduced on the Bt varieties, which allowed them to outyield the Sudanese controls by an average of 136 per cent under irrigation and 166 per cent under dryland growing conditions. Production costs were reduced by over 25 per cent.[75]

Approval was granted for commercial release in 2012. Over 20,000 hectares of Bt cotton were planted at four different locations, under both rainfed and irrigated growing conditions. Reports suggest they were planted by 10,000 smallholder farmers.[76] By 2016 this area had tripled, with more than 60,000 hectares under cultivation. The Ministry of Agriculture hopes to double this current production in the next few growing seasons.[77]

As in the case of Makhathini, current efforts to grow GM cotton follow many of the same principles as during the era of colonial-sponsored cotton production: they consist of a technocratic, intensive push emerging out of top-down authoritarian rule, with little emphasis on farmer input. A longer-term history of pest control efforts reveals an enduring battle against a variety of pests including white fly, jassids, aphids, thrips, termites, stainers, and moths.[78] The advantage of Bt control of bollworm is a continuation of a compartmentalized approach that promotes one solution to one pest without taking the others into account. Not surprisingly, a global comparison reveals that the singular focus of Bt cotton on bollworm control is more successful in environments where bollworm is the predominant pest, such as China or Mexico.[79] Already, Sudanese officials have identified Bt cotton's susceptibility to other pests and diseases, particularly the leaf-hopping pest jassid, as one of the technology's major downsides.[80] It is an open question whether growing Bt cotton makes sense given the high diversity of insect pests in Sudan.

Declines in cotton quality have also been observed in Sudan, echoing Burkina Faso's experience with Bt cotton. Cotton experts within the Ministry of Agriculture recount that prices received on the international market for Bt cotton were lower than for the traditional long-strand Sudanese varieties, and that domestic agricultural research centres are investing in efforts to rectify this dip in quality. Still, overall increases in aggregate production have supplemented lost income due to lower prices, and Sudan is actively exploring new ways to use Bt cotton domestically, including spinning, weaving, and the production of cottonseed oil. At present, cotton officials remain confident that their trajectory with Bt cotton will not follow the Burkinabè example: "We are not very interested in the experience of Burkina Faso because every research experiment differs from one other, and Burkina Faso plants cotton with rain whereas most of the planting in Sudan depends on irrigation from the river."[81]

More foundationally, the overreliance on technical solutions such as pesticide sprays has undermined the use of cultural methods that prove extremely valuable in controlling insect outbreaks, including the planting of trap crops, weeding, crop rotation, and the uprooting of cotton stalks at the end of each growing season.[82] The introduction of a new technical solution such as Bt without a concomitant emphasis on the broader spectrum of pest control practices seems unlikely to produce different results than the technocratic efforts that preceded it.

MODELLING THE POTENTIAL OF BT IN OTHER AFRICAN COUNTRIES

South Africa, Burkina Faso, and Sudan remain the only countries on the continent to have commercialized Bt cotton. A number of recent studies have emerged that offer concrete predictions of the potential benefits that will accrue to African countries should they decide to adopt this technology. These ex ante studies – that is, predictive studies that evaluate the potential for this new breeding technology to benefit farmers before it has been commercially released – have relied primarily on economic modelling to measure the changes that Bt cotton would precipitate.

The primary method used to assess these benefits is the economic model, mathematical constructs that represent economic processes by relying on a set of variables assessing the relationships among them. At their essence, economic models are theoretical tools that simplify complex processes in order to better understand how they function. A number of different models have been used within agricultural economics to evaluate the potential costs and benefits of Bt cotton adoption in Africa. The favoured approach is the general equilibrium model, a "bottom-up" type of economic model that begins with individual markets and agents and uses these to estimate the impacts of future technological introductions. These models have allowed researchers to calculate the financial losses that will accrue should African countries fail to adopt Bt cotton: on the lower end these losses have been calculated at US$13 million annually across the continent.[83] Higher-end estimates suggest that losses could be as high as US$41 million among seven major cotton-producing countries, or as high as US$80 million annually among French West Central African countries alone.[84] Other variations such as linear programming models have quantified the potential benefits that could accrue to countries that decide to adopt Bt cotton, with high-range estimates exceeding US$50 million per year.[85]

These ex ante modelling studies have been used to quantify the potential benefits that could accrue to individual countries willing to move forward with adoption. Egypt consistently figures as the country that would benefit the most from the introduction of Bt cotton, with estimated returns that are more than four times greater than other cotton-producing countries south of the Sahara.[86] A partnership between Monsanto and the Egyptian government led to the establishment of an experimental program designed to introgress Monsanto's Bollgard II technology into local Egyptian germplasm.[87] Five long-staple-length GM varieties were planted in two locations for CFTs in 2007. In 2009, the NBC authorized the production of basic seed for eventual commercialization. But efforts have since stalled.[88] While the reasons for this abrupt stop are disputed, three informants with knowledge of this partnership suggest that progress was derailed primarily by the upheaval that accompanied the Arab Spring.[89]

Kenya is another country projected by economic models to benefit significantly from the adoption of Bt cotton. The cotton sector in Kenya has declined precipitously from its peak in the 1970s and 1980s. Ginneries were closed across the country. Farmers in western Kenya shifted away from cotton to rice, while farmers in eastern Kenya switched over to maize and pigeon pea.[90] Proponents argue that Bt cotton offers the most promising means for reviving this industry, by reducing the high production costs, which remain the main barrier for many producers: pest control can account for as much as 32 per cent of a farmer's total production costs.[91] One recent report produced by the Kenyan government suggests that mass production of GM cotton would create 50,000 jobs and generate more than US$200 million in annual revenues.[92]

Bollworm is a particularly acute problem in the cotton-growing areas of Kenya, where the pest can destroy the entire crop if it is left unchecked. The agronomist in charge of Kenya's Bt cotton experiments suggests that the enhanced pest control offered by Bt cotton will allow Kenyan farmers to increase their gross margins by more than KSh 20,000.[93] But commercial release has stalled due to the 2012 cabinet ban, despite the completion of the scheduled experimental trials and its readiness for release in 2015. Efforts are underway to boost support for GM cotton and reverse the long-standing ban. A study was commissioned to model the potential benefits to cotton growers in Kenya – though Kenya ranked near last among the major cotton-producing countries in East and North Africa in terms of potential benefits. Proponents of genetic modification have also enlisted supportive politicians to help tout the benefits of Bt cotton. A National Stakeholders Forum was convened in 2014 to allow impacted politicians an opportunity to voice their opposition to the ban. The governor of Kisumu County – in the far west of the country on the coast of Lake Victoria – recently lamented the ban's impact on cotton producers in his district, arguing that Bt cotton was needed to make cotton a viable prospect for local producers.[94]

Experiments with Bt cotton are also currently underway in Ghana. While cotton was a key part of President Kwame Nkrumah's plan for industrialization and agricultural development in the 1960s and 1970s, production levels never managed to exceed 20,000 bales due to a combination of low producer prices, inconsistent services and late payments.[95] By the time the cotton sector was liberalized in 1985 there were approximately 50,000 producers in the country, concentrated almost exclusively in the north. Production surged during the late 1990s, peaking at 87,000 bales in the year 2000.[96] But the newly operating private cotton companies, all of which were local, struggled with the twin challenges of sustaining farmer commitment to cotton production and recovering credit for input delivery. A failed move towards exclusive zoning rights, combined with poor growing seasons early in the new millennium, wiped out the final remains of formal production.[97] Today cotton production remains a virtual "ghost industry."[98]

As in Kenya, Bt cotton is being touted as a means of reversing the textile industry's long decline and offering a return to its former glory. Ghana started its multi-locational trials with cotton in 2013. This is notable because it was this was the first example of an African nation choosing to forgo the CFTs that are generally mandated as the first step in the experimental approval for a new GM variety. In this case, Ghanaian officials felt confident that the extensive testing undertaken just across the northern border in Burkina Faso had been sufficient and did not require these safety measures to be replicated. So the first Bt cotton planted in Ghana was done so in farmers' fields, without restrictive fencing, in multiple agro-ecological zones.[99]

The results from experiments were largely promising. The scientist in charge of the experimental trials confirms that farmers were reporting higher yields and lower pesticide applications (these declined from six to two for most farmers).[100] In 2010 the Ghanaian government launched its National Cotton Revival Strategy, supported by the World Bank, the International Finance Corporation, and the United Nations Industrial Development Organization, which involved a radical restructuring of the industry with the aim of attracting international private partners. In April 2015, the minister of agriculture constituted a Cotton Development Authority, with a mandate for "revamping of the cotton industry to strengthen the promotion of private investment in the sector."[101] Bt is an important component of this broader strategy, viewed by proponents as "having the potential for solving the production issues." This is evidenced by the selection of the chairman of the new Cotton Development Organization (CDO), who previously served as director general of the Council for Scientific and Industrial Research (CSIR), who himself has been based at the research station coordinating the Bt cotton trials.[102] So Bt remains a key part of this more holistic approach designed to restructure the entire cotton value chain, including supporting marketing efforts and developing the capacity of domestic buyers.[103]

The unpredictable trajectories of Bt cotton in Egypt, Kenya, and Ghana challenge the quixotic projections embedded within these economic models. Most models rely on crucial assumptions that rarely hold true in practice. For instance, many assume a linear relationship between Bt cotton and its impacts on labour, yields, and input use across countries (an assumption known as Hicks neutrality). Also, most models are predicated on 100 per cent adoption rates (which never occurs in practice) and that all cotton is produced for fibre (which is not the case in most African countries, where substantial amounts of cottonseed are produced as animal feed).[104] Even when these assumptions are acknowledged and corrected for, such meta-level models struggle to take into account significant variation in climate, geography, farming practice, household decision-making, and institutional politics. As we have seen in the cases of South Africa, Burkina Faso, and Sudan, these context-specific variables are crucial in determining farmer outcomes with Bt cotton. Surveying the series of countries with experimental programs dedicated to GM cotton underscores the gap between these economic representations of GM cotton's potential and the realities faced by farmers there.

BT COTTON'S PROSPECTS IN UGANDA

Will the optimistic scenarios propagated by these economic models translate into benefits for smallholder farmers? I set out to answer this question in Uganda, a country proponents argue is ideally positioned to capitalize on the ecological and economic advantages of Bt cotton. Uganda's cotton industry was one of the continent's most vibrant throughout the twentieth century. The British attempted cotton cultivation in most of its eastern and southern colonial possessions, and Uganda's combination of heavy sun exposure and regular rains made it one of the most promising sites. In 1942, the Empire Cotton Growing Corporation consolidated its entire breeding operations at a central station at Namulonge, about twenty kilometres north of Kampala, strategically located in the region with the greatest growing potential.[105] This served as the hub for experimentation and dissemination efforts throughout the continent.

Domestic production thrived in the decades that followed (see figure 3.2). Cotton emerged as the nation's most important commodity crop and Uganda's cotton industry ballooned to become the third largest on the continent, after Egypt and Sudan. But production collapsed in the early 1970s: from a peak of 84,000 tons of cotton produced in 1969 to under 5,000 tons produced in 1979.[106] A variety of factors contributed to this precipitous decline, but the two most important were the series of oppressive regimes and widespread civil war that characterized the 1970s and 1980s, in combination with shortcomings within the industry itself, especially the low quality of cotton produced, the dearth of domestic buyers, and difficulty accessing rural credit.[107]

The end of the civil war in 1986 ushered in a period of reinvestment and reform designed to revive the cotton industry. Spearheaded by the World Bank, these efforts were aimed at stimulating research,

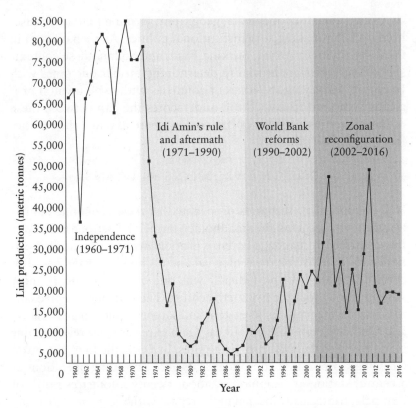

Figure 3.2 Cotton lint production in Uganda, 1960–2016. Original figure based on data from Cotton Development Organisation (CDO), "Uganda's Cotton Sub-Sector at a Glance," http://www.cdouga.org/production/production-trends-earnings; and the Food and Agriculture Organization of the United Nations (FAO), FAOSTAT, http://faostat3.fao.org.

disseminating improved varieties of seed, liberalizing marketing, and revitalizing the system of processing and ginning that had been deci-mated by decades of war.[108] This period of reinvestment culminated with the establishment of the CDO in 1994, which was inaugurated as the focal point for cotton production in the country, including coordinating, regulating, and monitoring all cotton activities. The CDO began registering participants from all sectors and implemented standards for seed cotton and lint moving then to mandating prices for seed cotton and providing seed to farmers.[109]

In spite of these major investments cotton output continued to lag, hovering around 20,000 tons during the late 1990s.[110] This time, experts identified insufficient use of inputs, especially pesticides and fertilizers, as the primary obstacle to production. A rash of new ventures, facilitated primarily by ginners, sought to provide these inputs to farmers as credit. None were successful. In response to these disappointing yields, the CDO restructured the industry into zones in 2003, designed to allow ginneries protected access to growers and enable contributions to a centrally administered fund that would dole out subsidized seeds and inputs to farmers. Again, there was an initial burst in production levels but these fizzled out soon after. Today, cotton remains the primary commodity crop for between 200,000 and 300,000 low-income households across the country, almost all of which cultivate cotton as a commodity crop alongside their subsistence staples.[111] Most growers put no more than half a hectare under cotton.[112] Average yields remain among the lowest on the continent.[113] At the national level, current production hovers around 22,500 tons, which proponents argue is less than 15 per cent of the country's full potential, estimated to be between 150,000 and 185,000 tons annually.[114]

By the turn of the millennium, GM cotton had emerged as the most promising means of overcoming these varied constraints related to stagnating production levels. The particulars of how the relationship between Monsanto and the government of Uganda originated are disputed. Monsanto claims that it was initially approached by the Ugandan cotton ginners' association in 1999, which viewed the company's patented Bt and Roundup Ready cotton as particularly well suited to Uganda, given the country's persistent problems with pests and weeds.[115] However, the then chairman of Uganda's CDO suggests that it was Monsanto who made the first overture in 2002 and that they "pushed and pushed" GM cotton as a preferred solution.[116] What is known for certain is that a national task force was created to investigate the possibility of testing Monsanto's GM cotton in 2003. This task force recommended that Uganda move ahead with experimental trials, with the aim of having a farm-ready product within five years.[117]

In August 2007 the Ugandan NBC issued permits for two CFT sites to investigate the potential of Monsanto's insect-resistant and herbicide-tolerant cotton varieties.[118] Two of NARO's regional sites were selected: the Serere station, in the semi-arid east, is mandated with cotton research and was chosen as the primary site; a secondary site was established in Kasese to reflect the higher elevations and precipitation levels facing western growers. Both CFT sites initiated planting in 2009, containing both Bollgard II (Bt cotton) and Roundup Ready Flex.[119] In the second year of the trial (2010), the stacked variety was added to each of the CFTs.

The first two years of planting were deemed a success. While no official data were published, informal reports and media coverage suggested that all three varieties demonstrated their traits effectively. The NARO scientist in charge of the trial reported that Bt "worked effectively against bollworm," while herbicide-tolerant (HT) cotton withstood Roundup Ready sprays "while weeds were wiped out."[120] Preliminary results were so encouraging that multiple visits were undertaken in 2009 and 2010 to bring relevant stakeholders to visit the CFT sites in person – including farmer leaders, media, and members of parliament – so they could see for themselves the experimental progress.[121]

But progress stalled soon after. The third plantings of CFTs, scheduled for spring 2011, were postponed. A minor scandal erupted after an inspection by the NBC reported that harvested GM cotton had been left out in the open, violating regulatory protocols. The accounting for this incident and restructuring of relevant protocols resulted in the trials being delayed by a full calendar year.[122]

Then, in June 2012, the CFTs were abruptly shut down. Monsanto's Africa office made the formal decision to suspend the CFT program. According to the Monsanto employee in charge of these trials, this decision to pull out reflected the company's frustration over the lack of progress with the country's biosafety bill, complaining that the science "was moving much, much faster than the legislation." He elaborated: "we realized we did not have an enabling environment for the technology to go on. That is, basically what we are talking about is the law. We did not have the law. And, the assessment was, why do we continue putting these technologies in confinement, yet we are supposed to be moving them further, to open field trials."[123] This employee viewed Monsanto's pull out as a strategic move, a means of putting pressure on the Ugandan government to accelerate passage of the biosafety bill, at which point Monsanto would return and pick up where it left off with the CFTs. This view was confirmed by Monsanto's former executive vice president and chief technology officer, who told a Ugandan reporter in 2013 that the decision to pull Bt cotton from Uganda stemmed from the unfavourable regulatory environment, and that the company would consider coming back to the country if the biosafety bill was passed.[124]

Despite these hurdles, proponents remain convinced that Bt cotton offers the most promising means for overcoming the stubbornly low levels of cotton production in Uganda. The two most comprehensive accounts of the challenges facing Uganda's sector – the first prepared by a senior agricultural economist at the World Bank, the second by a trio of economists at the International Food Policy Research Institute (IFPRI), at the request of the NBC – both recommend GM cotton as the single most effective mechanism for overcoming the industry's long period of under-production. The former concludes, "it is therefore imperative that Uganda ... accelerate their efforts to adopt GM technology,"[125] while the latter's findings "suggest that GM cotton has the potential of contributing to the improvement of cotton productivity in Uganda."[126]

My own research with Ugandan cotton farmers disputes these optimistic assessments. In 2013 I initiated a project that consulted with farmers across all the cotton growing areas of Uganda in order to assess their views on the potential advantages and limitations of GM cotton. We travelled to all corners of the country, reaching out to farmers in each of the major growing regions, including Arua District in the northwest corner of the country, Kaliro District in the east, Kasese District in the west, Lira District in the north, and Nakasongola District in the central region. In total we convened focus groups with eighty-six cotton farmers and conducted semistructured interviews with an additional two dozen sector representatives including ginners, extension workers, and cotton buyers, as well as officials from the CDO, the NARO, and the technology purveyor, Monsanto.

The information we gleaned from farmers underlines the complex interaction of ecological, political, and institutional obstacles to expanding cotton production in Uganda, and the limitations of relying exclusively on improved seed to overcome these. Since the farmers who participated in our study were sampled purposively, rather than randomly, we cannot suggest that this particular subset is representative of farmers everywhere. On the whole, though, their basic characteristics line up with those described elsewhere: typical cotton farmers own small plots, have little access to inputs such as pesticides and fertilizers, and produce relatively low yields.[127]

The question we sought to answer was whether increased productivity associated with GM cotton would translate into increased yields for farmers, enhancing their profitability. As we saw in the case of South Africa and Burkina Faso, farmers will choose to adopt GM cotton if the additional costs associated with its production (namely the technology fee and the increased requirement for fertilizers and other inputs) are outweighed by the yield-enhancing benefits of reduced pest damage and/or savings from reduced pesticide use.

Let's deal with the potential benefits from reduced pesticide exposure first. On average, the cotton growers who participated in our study sprayed three times per season, spending an average of just under US$5 per acre per spray. At first glance, these high cost burdens imply significant potential savings, since adopting Bt cotton will lower pesticide burdens and decrease costs. But this direct link between reduced bollworm damage and increased pesticide savings fails to fully appreciate the complex set of pest management challenges facing Ugandan cotton growers. The major cotton pests in Uganda fall into three groups: first, the sucking pests such as jassids, aphids, and lygus, which attack in the weeks following germination; then spiny, American, and red bollworm, which attack when the flower transforms into buds and then bolls; and finally stainers, which suck the cottonseed and ruin the lint. Farmers in our project were split as to which pests were the most damaging. Overall, farmers identified lygus as the most damaging pest, followed by aphids, bollworm, and then jassid. Regardless of their order of priority, any intervention designed to increase cotton yields will need to take into account all three categories of pest. One of the country's leading cotton experts explained the ideal pest management strategy as follows: first, applying a seed dressing chemical to Bt cotton that coats the seed to inhibit early infestations of aphids and lygus (inserted into the local Bukalasa Pedigree Albar variety that is already hairy to inhibit jassid) and to manage bollworm; and finally one or two sprays of organophosphate towards the end of the season to guard against stainers and late infestations of aphids or lygus.[128] Such a comprehensive strategy would depend on significant credit availability (to ensure farmers could afford the more expensive seed dressings and Bt seed) as well as constant vigilance and scouting to ensure timely application of late pesticide sprays. It requires a complete overhaul of the current pest management system. Bt cotton is one small part of the broader pest management challenge: on its own, mitigating bollworm damage via resistant varieties will have little impact on farmers' bottom line.

Even the potential savings that would accrue to farmers who adopt insect-resistant cotton is up for debate. Proponents portray farmers' use of pesticide applications as largely ineffectual and wasteful, which implies significant cost savings in using a GM variety.[129] But other experts argue that farmers' use of pesticides is both efficient and effective. The former head of the CDO argues that many farmers are able to control bollworm with only two sprayings of pesticides, which together comprise only a minor expense to the farmer.[130] The World Bank report corroborates this view, showing that spending on pesticides accounts for "only a small share of the total cash expenses" incurred by cotton growers, leaving little room for meaningful savings.[131]

How about potential yield increases resulting from lower damage due to pests and weeds? In the most comprehensive study to date, IFPRI researchers have estimated a potential yield increase of 28 per cent from adopting Bt cotton and 29 per cent from adopting HT cotton. The figures are modelled off of their own farmer survey in two separate growing areas, in which participants report that 76 per cent of losses are due to bollworm. Our own conversations with farmers across the five main growing areas produced very different results: bollworm was ranked as only the third most damaging pest, after lygus and aphids. Indeed, the modellers themselves acknowledge that their own estimate of bollworm damage is likely inflated, noting that "[farmers'] perceptions are usually upward biased, given that is rather difficult for farmers to isolate the net effect of one constraint from all the other constraints they face."[132] It seems likely that this figure of 76 per cent overestimates the degree to which bollworm causes cotton losses, and underestimates the degree that the cocktail of other major pests account for these losses. Given this, these estimates for potential yield increases seem farfetched.

Another consideration that remains underappreciated within this debate is the incongruity between the monoculture production system required for GM cotton and the intercropping system that is practiced by the vast majority of farmers. Nearly all of the farmers who participated in our study intercropped their cotton, most with some combination of maize, beans, and groundnuts. Farmers at Nakasongola in the central region explained to us that intercropping subsistence crops alongside cotton was the only way they could make this enterprise viable, as the secondary crops act as a cushion against the highly fluctuating prices for cotton. Farmers in the northern regions – especially women farmers – relayed that intercropping acted as insurance against potential low cotton yields, offering guaranteed food that they could take home or use to pay labourers if yields were disappointing. Adopting GM cotton would force growers to plant cotton without this cushion.[133]

Our conversations with farmers further revealed a number of structural factors that complicate the simplistic notion that integrating Bt or HT cotton will increase yields end livelihoods. One set of potential hurdles relates to the GM variety that has been used throughout the experimental process. The CDO distributes only one type of long-staple-length seed, Bukalasa Pedigree Albar, known commonly as BPA, which originated from the Empire Cotton Growing Corporation's headquarters in Uganda. BPA is well known for its quality and staple length, and for being well suited to the ecological conditions in Uganda. In particular, this variety was bred for its "hairiness" – that is, long hairs on the underside of the leaf that deter infestations from leaf-hopping insects such as jassids. The CFTS initiated in 2009 use an American variety, SureGrow 125, which is non-hairy. Unsurprisingly, this variety proved extremely vulnerable to jassid infestations. The Ugandan scientists and policy officials were unanimous in recommending that Monsanto follow the same protocols initiated in Burkina Faso, in which the Bt event was inserted into the local germplasm. There has been no official response from Monsanto as to whether they are willing to undertake this costly and time-consuming introgression. When asked about this, officials were noncommittal, responding that with the CFTS they were "looking mainly at the technology, we are not looking mainly at the variety."[134] Monsanto's commitment to an experimental program using an American cotton variety suggests a certain reluctance, at least initially, towards integrating GM events within local varieties. As we have seen in the case of Burkina Faso, ensuring the persistence of the favourable traits expressed by the local variety is crucial to the success of genetic modification.

A second issue related to seed is the distinction between determinate and indeterminate flowering. The former produces cotton bolls that mature at around the same time, which is preferable for mechanical picking, while the latter produces mature bolls in staggered fashion, which is preferable for handpicking. Not surprisingly, then, the American SureGrow variety planted in CFTS is determinate, while the local BPA variety is indeterminate. The majority of farmers in our study (57 per cent) prefer indeterminate varieties because it allows them to spread the hard labour of picking over a longer period, and gives them more opportunity to take advantage of the school holiday in January and February, which frees up family labour to work on harvesting. As one farmer in Kaliro District put it: "I have two wives and each of them demands for money at different times so it is good if there is some harvest that can be made at different times." Farmers also relayed that the longer growing season associated with indeterminate varieties allowed them to take advantage of favourable changes in market price. Other farmers expressed concern over determinate varieties that might overwhelm their limited capacity to handpick ripe cotton bolls. Thus, the concern around GM cotton that is not backcrossed into local varieties is that it will produce harvesting patterns that are incompatible with the needs and realities of the country's smallholder growers.

Perhaps the most significant barrier to successful adoption is the final one: the technology fee. Horna, Zambrano, and Falck-Zepeda produce two scenarios, one in which GM cotton is given for free and one in which a technology fee is charged. The different outcomes for farmers are substantial: with the technology fee included in the model the profitability a farmer is expected to gain from introducing genetic modification technology drops more than four-fold – from US$169.33 per hectare to US$42.91 per hectare. The benefit-cost ratio is cut by more than a third, and the downside risk increases from 27 to 44 per cent when this technology is taken into account.[135] It is highly unlikely that farmers would be able to see any financial benefit from GM cotton once the added cost of the technology fee is taken into account. Even proponents of GM cotton recognize that this added cost represents a major obstacle to successful adoption: once these costs are added it is just the wealthiest subset of cotton farmers who would see any benefit.[136] These researchers conclude that financial costs represent a principal hurdle to adoption: "the technology fee is a critical determinant of both the level of benefits to the society and the downside risk to producers."[137]

This hurdle of the high technology fee underlines the unfavourable financial constraints facing Ugandan growers. Farmers emphasized that the most important variable influencing both their decisions on whether to plant cotton and how much cotton to plant was the price they received from the buyer: this was consistently identified by farmers in ranking exercises as the most important factor in deciding whether to plant cotton. Farmers explained that cotton was such a labour-intensive crop that only a price received above 33 cents (USD) per kilo was sufficient to make this a profitable enterprise. A male farmer in Arua District put in the following terms: "price is the only factor against which our decision to continue growing cotton is based; the price inspires us to grow more cotton." Another farmer in Lira District echoed this sentiment: "price is the only factor that can compensate for the farmer's time, cotton takes a lot of time in the field." A female farmer in Kaliro District agreed: "price of buyer is important because it is useless to grow much when the price is very bad ... price is the biggest incentive in growing cotton." Introducing GM cotton will not influence the price received by growers, which primarily reflects global market conditions. As such, it fails to respond to the farmers' most urgent need.

Price looms large for farmers who struggle to make cotton growing a revenue-generating enterprise. Just under half of the farmers in our sample rent their land from cotton buyers, who mandate that rent be paid in kilos of cotton (on average the rent expected was around ninety kilograms of cotton, equivalent to around US$40 at the time of our visit). This accounts for just over 50 per cent of the cotton produced per acre. One male farmer from Bumanya Subcountry in Kaliro District explains: "the buyers are willing to rent land to us because then they give us the pre-condition to sell to them. You always have to sell to your landlord, so he has more to sell to the ginnery." Farmers noted that low prices and restricted markets squeezed their options – one farmer in Nakasongola lamented: "the monopolistic buyers set prices which are not sufficient for us as farmers." The World Bank confirms that price is the single most important variable impacting cotton profitability: "when prices change, farmers act accordingly."[138]

As with the case of Makhathini and Burkina Faso, the push towards GM cotton in Uganda is likely to produce very different outcomes for distinct actors within the sector. The two dozen industry representatives whom we interviewed in our study – mostly ginners and cotton buyers – were unanimous in supporting GM cotton as a means of spurring stagnating production levels. Some industry insiders harboured more transformational aspirations, suggesting that GM cotton might encourage independent producers to band together into block farmers.[139] Within this view, GM cotton is touted as a means of raising overall production levels to help fill the capacity of existing infrastructure: the most recent estimate available suggests that over 70 per cent of ginneries in the country are operating below capacity.[140]

At the same time, there are many factors that make the GM cotton currently under consideration incongruent with the realities facing smallholder Ugandan cotton producers. Questions regarding the appropriateness of the seed, the accompanying growing system, and its added cost raise doubts around the potential for GM varieties to generate profits for smallholder growers.

Proponents suggest that Uganda's cotton renaissance depends primarily on "the availability of good quality seed."[141] But in reality farmer preferences – and the cotton industry more generally – are much more complex. The transition to GM cotton would have wide-ranging implications for the nation's cotton producers, impacting cropping patterns, pest management strategies, and seed cost, and yet have no impact on the single most important variable identified by farmers in our study: the price received.

CONCLUSION

The dominant portrayal of Bt cotton in Africa has been one of promise and success. The examples of the Makhathini Flats in South Africa and Burkina Faso demonstrated the potential for Bt cotton to help smallholder farmers in Africa. This chapter has exposed the gap between this rhetoric and the reality on the ground. Makhathini's early success was due to structural factors such as credit availability and preferential market access. But these favourable conditions could not be sustained over time. Once this enabling context disappeared, so too did the adoption rates. The vertical control of the Burkina Faso cotton industry propped up Bt cotton for over a decade. But when issues of inferior quality emerged, this same centralized apparatus instigated a full reversal and Bt cotton was phased out over three years. Taken together, these two cases underline the inextricability of technology and context. As Melinda Smale and her colleagues emphasize, "the institutional and social context of technology introduction is often of greater significance for determining the direction and magnitude of impacts than the effectiveness of any particular trait."[142]

Sudan's recent experiences with Bt cotton underline the particular kind of context the technology needs in order to succeed: large-scale application, heavy mechanization, monoculture farming, centre-pivot irrigation, and state sponsorship. If this formula is maintained then there is a good chance it will prove economical over time, though it is likely that secondary pests will eat into these profits.[143] But it is naive to think that this model can be successfully transplanted to smallholder growers in Uganda, Kenya, Ghana, or elsewhere in Africa. Although forward-looking studies using econometric modelling or large-scale surveys predict significant benefits to smallholder growers, incorporating evidence from farm-level studies underscores just how wide the gap is between the potential of Bt cotton and the practical benefits if offers to smallholders. To borrow from Dominic Glover, the portrayal of Bt cotton's success in Africa reflects "methodological weaknesses and presentational flaws."[144]

4

GM Maize in South Africa and Egypt

Maize is the most important carbohydrate staple across Africa. While not indigenous to the continent, maize now covers more land across Africa than any other crop, accounting for just less than thirty-five million hectares harvested in 2013.[1] It is also the most widely consumed: maize represents almost a quarter of the continent's daily carbohydrate intake, making it the largest single source of calories across Africa.

Maize's prominence within African agriculture means it occupies a central position within debates around Africa's lagging levels of agricultural production. In his 2015 annual letter to his philanthropic foundation, Bill Gates uses the case of maize to explain the burden of Africa's persistent and pernicious yield gap. He laments the fact that African maize yields are only a fraction of what they are in more sophisticated farming systems: yields of American maize are more than five times that obtained by the average African farmer. Gates then proceeds to lay out his big bet for fixing these discrepancies: "The world has already developed better fertilizer and crops that are more productive, nutritious, and drought- and disease-resistant; with access to these and other existing technologies, African farmers could theoretically double their yields." The BMGF is aiming to increase productivity by more than 50 per cent, boosting average maize yields from 32.6 bushels per acre today to 48.9 bushels per acre in fifteen years. Gates is optimistic that this investment in increasing productivity "will make a world of difference for millions of farmers."[2]

New breeding technologies are a key element of this pledge to revolutionize maize production in Africa, and the BMGF has been at the forefront of agricultural programming designed to increase maize productivity. Much of this investment has been funnelled to CIMMYT, the CGIAR node for maize improvement efforts, which has received more than $180 million from the BMGF as of 2014. In addition, the BMGF has directly invested tens of millions of dollars in the Water Efficient Maize for Africa (WEMA) program (discussed in chapter 6).[3]

Current efforts to enhance maize breeding efforts across Africa revolve around three distinct but closely related GM versions, all patented and owned by Monsanto: Bt maize (so called because it is genetically modified with a gene from *Bacillus thuringiensis* to promote insect resistance), HT maize (so called because it is genetically modified to promote herbicide tolerance to Monsanto's Roundup Ready herbicide), and BR maize (also known as the stacked version, so called because it combines the insect resistance from Bt with the herbicide tolerance to Roundup Ready). This chapter surveys existing efforts to integrate GM maize within African agricultural systems. It pays particular attention to the case of South Africa (the first country to commercialize this technology) and Egypt (the only other African country to commercialize this technology). It concludes with reflections on the incompatibilities between the technology and the context of smallholder maize production in Africa, arguing that any benefit accruing from this technology will only materialize once the gap between these two realms is overcome.

A (SHORT) HISTORY OF MAIZE
AND MAIZE BREEDING IN AFRICA

Maize is an open-pollinating species whereby genetic material is exchanged via wind and insects, making it possible for it to be fertilized by itself (self-pollination) or by its neighbours (cross-pollination). If these flows of pollen are left uncontrolled, each individual maize plant will express unique genetic and physiological characteristics, both from the preceding generation and from each other. When maize self-pollinates, the resulting progeny tend to exhibit characteristics inferior to their parents. But when it cross-pollinates the resulting progeny tend to outperform their parents. This process is known as hybrid vigour.

This high rate of genetic variation rapidly produces individuals that express beneficial traits such as high yields or drought resistance, as well as others that express detrimental traits such as susceptibility to pests or weakened stalks. Conventional maize breeding focuses on identifying and perpetuating these high performers and eliminating weaker ones. Two types of improved varieties are most common. The first kind of improved seed is derived from open-pollinated varieties (OPVs), which refers to genetically diverse populations of maize that have been purposely bred to express preferential traits within a diverse set of production conditions. The major advantages of OPVs are their broad adaptability and stability amid variations in growing conditions. These can also be reused without deterioration in desired characteristics. The second kind of improved seeds are hybrids, whereby carefully controlled pollination produces homogenous varieties that demonstrate beneficial traits. The inevitability of cross-pollination, though, means that these beneficial traits are diluted in subsequent generations, making seed replanting and recycling less advantageous. In order to maintain these beneficial traits, farmers must buy new hybrid seeds every year. Hybrids tend to outyield OPVs due to an accumulation of dominant favourable genes, though the latter tend to be favoured by smallholders due to the high levels of stability over time, broad adaptability to local growing conditions, lower cost, and ability to replant without the loss of beneficial traits.

Despite its widespread prevalence across the continent, maize is not native to Africa: it was imported from the Americas sometime in the sixteenth century, rapidly expanding across Africa as it displaced previous carbohydrate staples such as sorghum and millet. Its genetic diversity and malleable nature facilitated its spread across Africa's diverse environments, penetrating into highland plateaus, humid tropics, and semiarid substropics.[4] Yellow maize emerged as the primary feed for animal consumption, while white maize became the major food staple across much of the continent.

The history of maize breeding on the continent is one in which the relative investments in OPVs and hybrids reflect the political priorities of governing regimes that prioritized the few over the many. Throughout the twentieth century, hybrid development in southern and eastern Africa was explicitly designed to boost production levels on large-scale settler-owned farms, which benefited from favourable climatic conditions, abundant labour, and heavy investment, and insulated them against competition from African smallholders who were predominantly located on plots with less enviable growing conditions.[5]

The most successful hybrids, which emerged out of Zimbabwe (then Rhodesia) in the 1960s and 1970s, are what environmental historian

James McCann characterizes as high-yield/high-management varieties. These were capable of obtaining yields upwards of 40 per cent higher than the best available OPVs, under an intensive management regime that included mechanized tractors, heavy fertilizer use, and well-timed irrigation systems.[6] It was no accident that such conditions were well out of the reach of the average smallholder: the hybrid research program was focused on maximizing the potential of the prime soils controlled by white commercial farmers, with the aim of boosting supplies of maize flour for the rapidly growing urban centres in southern and eastern Africa. Maize breeding was structured to privilege settler production at the expense of African farmer output. In McCann's words:

> Government support for maize research was not just a commitment to agricultural modernization; it was part of a larger plan to ensure the economic and social base of white rule … the emphasis on the study of improved and hybrid maize took place at precisely the same time as the systematic attempts both before and during the federation to undermine African maize production and secure the valuable agricultural lands for the settler economy.[7]

McCann reveals how traits favoured within twentieth-century breeding programs reflected the political priorities of the day. Breeders privileged a longer growing season – each day's delay in planting translated into a 1 per cent or 2 per cent loss in yields – a challenge that was overcome by farmers with the capital or credit to use tractors or oxen to plow their fields early in the season, but which severely disadvantaged smallholders who relied primarily on hand tillage and tended to shy away from early planting due to labour constraints and crop rotations.[8] Other characteristics of high-yielding hybrids were similarly incompatible with the rhythms and realities of smallholder production, including softer endosperms that made them more susceptible to damage by insects when stored for long periods of time, and a long-standing bias on the part of breeders against varieties that could thrive in intercropping systems, which were favoured by farmers throughout eastern and southern Africa (particular in combination with African staples such as beans, sorghum, or cassava). And, of course, these hybrids suffered from dramatic losses of hybrid vigour – losses with replanting that often exceeded 50 per cent – meaning that farmers had to purchase new seeds with every growing season.[9]

Thus, the history of maize breeding in Africa is one in which, in McCann's words, the prioritization of white commercial farmers was embedded in the genetic makeup of these new high-performing varieties. Traits privileged by hybrid breeders throughout the decades – including higher yields, optimal performance under ideal growing conditions, and longer growing seasons – reflected the conditions and priorities of industrial agriculture. But these improved varieties were ill-suited to the ecological and economic realities facing African farmers. Heavy investment in hybrid maize and the corresponding lack of investment in OPVs created two divergent maize sectors: one dominated by wealthy settler growers who depended on hybrid seed, and the other dominated by small subsistence black growers who depended almost entirely on OPVs. Maize transitioned away from the priorities of those who cultivated it as a subsistence crop and emerged instead as an "industrial grain, and not a vegetable to round out rural diets."[10] Thus, maize breeding in the twentieth century was characterized by a deliberate shift away from maize as a subsistence crop cultivated by the many towards a commodity crop cultivated by the few.

SOUTH AFRICA: THE SUCCESS STORY

Nowhere were these asymmetrical breeding priorities more pronounced than in South Africa, where the comparatively late transition to democracy left it far behind neighbouring countries whose national research programs had transitioned to smallholder-centred research agendas decades earlier.[11] At the time South Africa achieved multiracial democracy in 1994, maize growing in the country was split sharply between the white, commercial growers who favoured hybrid seed and the black, smallholder growers who relied primarily on OPVs. This divide persisted even after the apartheid regime had been toppled: at the turn of the millennium 90 per cent of the country's more than two million smallholders growers planted OPVs or recycled seed bought in previous seasons.[12]

It was within this context of uneven seed access that the first GM maize variety was released at the end of the twentieth century. As with cotton, it was South Africa who would host the first experiments and the first commercial release of this new GM variety. The first GM maize ever planted on the continent was Monsanto's Bt (YieldGard) maize, which has been genetically modified to express the Cry1Ab gene, extracted from the soil microorganism *Bacillus thuringiensis*. Once ingested, this protein is fatal to larvae of the genus *Lepidoptera*. Internal tests of YieldGard, developed by Monsanto to control damage wrought by European stem borers, revealed suitable resistance to the two species of stem borers that cause the most damage in the South African context: the African stem borer (*Busseola fusca*) and the Chilo borer (*Chilo partellus*). These insects lay eggs between the leaf sheaths and feed on the plant's unfurled upper leaves. Combined, these two pests cause annual yield losses of around 10 per cent each year, which can translate into yearly losses of just under one million tons of maize.[13]

While GM maize was designed with commercial users in mind, proponents began to argue that the benefits associated with these technologies could be extended to poor farmers in sub-Saharan Africa (SSA), where pest infestations are even higher due to the absence of a dormant period and overlapping generations, where pesticides make up a higher proportion of total expenses, and where detection and monitoring are difficult and time-consuming. Supporters suggested that Bt maize was particularly well suited for small-scale farmers because it eliminated the considerable amounts of equipment and information required for proper insecticide applications, and reduced farmer exposure to harmful pesticides.[14]

GM versions of yellow maize – which accounted for approximately 30 per cent of South Africa's total production, consumed primarily as animal feed – were the first to receive approval, with commercial release in the 1998/99 growing season. Nearly all of these growers were large-scale operators: most held title or rental contracts for the land, and enjoyed relatively easy access to inputs, credit, extension services, and markets. Initial assessments of Bt yellow maize's performance were extremely positive: yield advantages hovered at around 11 per cent, while insecticide costs were more than halved.[15] Despite the increased costs associated with the technology – around 50 per cent more expensive than conventional seed – farmers reported a net increase in income across the board, averaging over US$27 per hectare. Within a few years somewhere between 50 per cent and 70 per cent of these farmers had adopted the technology, leading one exponent to proclaim that the "evidence from large-scale farmers indicates that Bt maize can increase yields, reduce pesticide use and give farmers substantial economic benefits."[16] Over the course of a decade, the increased yields that accrued to yellow maize growers were estimated to exceed 900 million rand.[17]

GM versions of white maize, which is produced primarily by smallholder growers for subsistence production, with excess often for sale at local markets, was released three years later in 2001. As with the rollouts of Bt cotton outlined in the previous chapter, the first introduction of Bt white maize in South Africa was carefully planned and protected. In 2001 Monsanto identified nine areas across four provinces where farmers cultivated smallholdings of maize without access to irrigation. They then invited 3,000 of these farmers to meetings convened in each area, explaining the advantages of Bt maize in the local language. Upon completing the workshops, farmers were given 250 grams of Bt maize seed alongside 250 grams of conventional seed (the extra costs associated with Bt seed, known as the technology fee, were waived in this first growing season). Farmers were encouraged to plant the GM version for free alongside its conventional counterpart. Estimates for both adoption rates and yield for these first few seasons were difficult to come by, due to the methodological challenges of comparing various pockets of adopters across the country and measuring yields on very small plots (generally under one hectare), where much of the maize produced is consumed in an immature form as green mealies, which hindered the ability of researchers to capture reliable estimates on yields.[18] Still, the overall portrait of these first few growing seasons was largely positive: average yield increases over the nearest conventionally bred equivalents were reported to be 32 per cent in 2001/02, 16 per cent in 2002/03 and 5 per cent in 2003/04.[19] Over the first six growing seasons, Bt maize yielded 12 per cent higher than conventional varieties due to reduced borer damage conferred by genetic modification. This research also recorded favourable farmer reviews of both the quality of the grain and the taste of the green mealies produced by the Bt seed. Acreage planted under Bt white maize jumped from 8.6 per cent of the total white maize in the area in 2004 to more than 50 per cent in 2006.[20]

Two new GM varieties followed suit in 2003/04 and 2007/08. The first was Monsanto's HT variety, which offered farmers a maize variety resistant to Monsanto's Roundup Ready herbicide, meaning that farmers could spray their fields with the herbicide that would kill all weeds but leave their maize stands unharmed; the second combined the two previous traits – insect resistance (Bt) and herbicide resistance (HT) in a stacked version (BR) that contained both traits. Both rollouts followed the pattern of the first and included demonstration plots, farmer days presentations, informational pamphlets, and supplemental training to government extension officers – all in the local language – provided by seed companies. Similar to the case of Bt maize, initial reports were promising. In 2006/07, one group of farmers in northern KwaZulu-Natal reported yields 184 per cent higher than conventional varieties, figures that were touted as evidence of the labour-saving properties of the HT-resistant varieties, which required considerably less labour for weeding.[21] These authors argue that total labour per hectare was lower for HT maize, due to the fact that most growers who planted GM maize did so in a no-till or low-till system, shifting their major labour expenditures from weeding (which is particularly onerous) to herbicide application (which is considerably less so). Their findings suggest that the 35 per cent fee increase that producers pay for HT seed was more than offset by a 42 per cent reduction in labour per hectare, resulting in a net benefit. These labour-saving properties are particularly beneficial for rural, small-scale producers who had been hit hard by a combination of events that combined to hollow out labour availability. The combination of widespread migration of younger males to urban centres in search of employment and the devastation of HIV/AIDS left many households headed by female and elderly farmers, who struggled to effectively control weeds due to limited time and energy.[22]

Exponents concluded that farmers understood the advantages conferred by herbicide resistance: by 2015, 50 per cent of GM seed sold to smallholders throughout South Africa was HT and 42 per cent was BR. The consensus that emerged from this series of on-farm surveys assessing the impacts of GM maize was a positive one: "The comparisons show that HT and BR maize produced under a minimum/no-till production system with a broad spectrum herbicide holds great potential for smallholder farmers due to improved weed control and especially its labour saving advantage."[23] The most recent figures suggest that HT and BR maize account for approximately 75 per cent of the total yellow maize crop, while Bt now accounts for only 25 per cent. These figures are lower for white maize, but not by much: HT and BR maize together account for 68 per cent of the total crop and Bt accounts for the remaining 32 per cent.[24] One recent study went further and disaggregated these benefits by gender, revealing that men are most impressed with high yields, while women farmers value corollary benefits like taste, quality, and ease of farming, especially as it reduced the amount of time that they (and their children) spent weeding.[25]

Just one decade after its initial release, the rise in acreage of GM crops across the country has been spectacular: from 3,000 hectares in 2000 to just under 500,000 hectares in 2005, to more than 1.5 million hectares in 2007, and more than 2.3 million hectares in 2013.[26] The most recent production figures available show that more than 84 per cent of all white maize and 91 per cent of all yellow maize planted in the country is now genetically modified.[27] One estimate suggests that the combined financial benefit that accrued to the nation's commercial farmers in the first decade of cultivation was just shy of two billion rand.[28] Taken together, this chronology of smallholder experiences with Bt, HT, and stacked varieties led proponents to conclude that these were beneficial technologies that could meaningfully impact both yields and livelihoods.

AN ALTERNATIVE ACCOUNT TO THE STORY OF SOUTH AFRICAN SUCCESS

As with the country's experience with GM cotton, there are two distinctive and at times conflicting narratives surrounding South Africa's experience with GM maize. The first is anchored by an unwavering adherence to the technology's success, both established and potential. The second seeks to contextualize these success claims by disaggregating some of the meta-level data presented above and interrogating some of the assumptions contained within it. For instance, the figure of 84 per cent of maize being genetically modified suggests that the overwhelming majority of farmers have committed to growing GM varieties. But without the necessary context, this figure is deceiving; it overrepresents the country's 9,000 large-scale farmers and their overwhelming preference for GM varieties of yellow maize, which serves primarily as animal feed.[29] In terms of the number of farmers adopting GM maize, its penetration is quite limited: only approximately 10,000 smallholder maize farmers planted GM maize in 2007/08, representing less than 0.5 per cent of the country's total of two million subsistence maize farmers.[30]

Why then have more farmers not decided to adopt these technologies, despite the numerous advantages surveyed in the previous section? The most illuminating analysis comes from a Swedish researcher, Klara Fischer (formerly Jacobson), whose doctoral research examined farmer responses to GM maize among smallholder farmers in the Eastern Cape province. Fischer's research exposes the significance of smallholder farmer responses to differences between aggregate and year-to-year yields. The data accumulated over the past fifteen years suggest that yields for all three GM versions of maize are higher than for both local OPVs and conventional hybrids, and that overall economic benefits are positive even after the increased technology fee is taken into account. But these aggregate totals mask sharp variations *between* seasons.[31] For both the Bt and HT varieties, seasonal fluctuations in precipitation produce variations in the potential losses from pests and weeds, and thus variability in the potential gains from investing in Bt and HT varieties.

In the case of Bt, this high level of variability stems from the ecology of the borers themselves: seasonal fluctuations in precipitation strongly influence both the prevalence and severity of infestations. The trend is towards higher infestations in seasons with heavier rains, as humidity is crucial to the moths' survival.[32] This high variability in borer infestations means that Bt yields will be much higher in years of high rainfall: in one such year, 2007/08, higher infestations meant that Bt maize outyielded conventional hybrids by more than 25 per cent. But it also means that in years of low rainfall farmers will have little need for this added protection, such as 2002 to 2005 when farmers in northern KwaZulu-Natal largely abandoned Bt maize, due to lower than usual levels of rainfall and a corresponding plunge in borer infestations, thus negating the need for Bt resistance.[33]

Proponents argue that these simultaneous dips in precipitation levels and borer infestations reflect abnormal conditions, and that the return of regular levels of precipitation will demonstrate both the ecological and economic benefits of Bt. But variability in precipitation is not abnormal in South Africa; rather, long-term studies of rainfall patterns reveal that these abnormal variations are indeed quite normal, especially in the eastern part of the country such as Mpumalanga, Eastern Cape, and KwaZulu-Natal, where smallholder producers dominate. The Climatology Research Group at the University of the Witwatersrand has identified a number of regular oscillation events that account for this periodicity, including shifts in the Intertropical Convergence Zone that correspond to changes in El Niño–Southern Oscillation events, as well as a fairly predictable eighteen-to-twenty-year oscillation.[34] Each of these oscillations accounts for as much as 20 to 30 per cent of the variance in summer rainfall, which has significant implications for maize growers across the eastern part of the country. Maize planting in South Africa generally occurs in October or November, with harvesting around May or June. There are two distinct rainfall periods that are crucial for the crop's growth: the first, extending from September to December, supports land preparation, planting, germination, vegetative growth, flowering, and pollination, while the second, spanning January through March, supports the cob-formation stage. Endemic levels of variation in precipitation is a predictable part of South African maize growing conditions, making it unlikely that the high inter-year variation in benefits accruing to smallholder farmers planting both Bt and HT maize will dissipate anytime soon.

Farmers have long recognized these links between precipitation levels and borer damage, and adjust their farming cycles in order to take advantages of favourable conditions. For instance, many smallholder maize farmers aim to plant their crops as late as possible, knowing that waiting out the first moth flight is one of the most effective mechanisms for avoiding borer damage.[35] Yet this practice is discouraged under Bt maize: according to one account, Monsanto materials dissuaded farmers from planting Bt maize late, suggesting that this makes the crop more vulnerable to borer attacks that might coincide with flowering, against which Bt offers little protection (Bt protects against young moths, but not adult ones).[36] This prescription for early planting is often resisted by small farmers: without sufficient rains, land preparation is much harder (particularly for farmers without mechanical tractors), while those who depend on oxen tend to wait as long as possible in order to allow the animals extra time to improve their stamina after the dry winter.[37]

The high levels of interannual variability built into Bt and HT maize technology is amplified by their increased costs. Farmers who choose to grow GM maize agree to pay a technology fee to the licence holder for every year that they plant GM seeds. While the price differentials have diminished over time, Bt and HT maize remain about 30 per cent more expensive than the nearest conventional equivalent, 80 per cent more expensive than the most popular conventional hybrid (known as PAN 6043), and as much as five times more expensive than most OPVs.[38] The increased cost and high levels of variability greatly augment the risk for farmers, who run the risk of paying extra for insect protection that they do not need in seasons with lower levels of infestations. In the 2003/04 growing season, for instance, farmers who opted for Bt seed were actually worse off – they were 12 per cent less efficient due to higher seed costs and marginal benefits.[39] Such dramatic swings in year-to-year productivity can be devastating for farmers with little economic buffer in the form of savings or surplus food stores, making it difficult for farmers to absorb these losses in the years when they have to pay extra for insect resistance that might not even be needed.

These insights resonate with the important findings offered by Daniela Soleri and David Cleveland, whose research with GM maize growers in Central and Latin America show the particular ways that smallholder farmers' perceptions of the benefits and risks associated with GM maize are distinct from their larger-scale counterparts. Soleri and Cleveland used participatory ranking exercises to evaluate farmer understandings of the advantages and disadvantages of GM maize among smallholders in Cuba, Guatemala, and Mexico. Farmers in their study were largely risk averse, preferring yield stability over higher yields, emphasizing the importance of meeting minimum thresholds in every growing season.[40] They found that farmers were skeptical of a number of the assumptions contained in breeding programs for genetic modification, including yields responsive to ideal environmental conditions and reliance on the formal crop improvement and distributions systems.

More generally, Soleri and Cleveland's research findings refute the two critical assumptions that underlie economic assertions surrounding the adoption of GM maize: first, that farmers will choose GM crops over other crop varieties because they offer the greatest utility in the form of the greatest profit, and second, that farmers are risk neutral profit maximizers.[41] They argue that the risk management process for smallholders is fundamentally distinct. Poorer farmers tend to plant fewer hectares of maize spread out over a greater area, and more varieties of maize, which makes these farms look very different than those of industrial agriculture. Increases in average yield gains despite some years with economic loss mean little to smallholder farmers who, unlike larger, more capital-intensive farmers, cannot spread gains and losses across seasons; thus resource-poor farmers tend to value the stability and predictability of the returns they get from technologies more than wealthier farmers.[42] Within this view, the yield patterns of GM maize – higher yielding with higher variance – are more appealing to larger, wealthier farmers than smaller, poorer farmers.

Thus, the issues of scale and wealth, which are difficult to separate out, fundamentally determine how the ecological and economic benefits of GM maize play out for farmers. Sixty per cent of largeholder yellow maize farmers spray with pesticides to control the stalk borer, so it follows that the majority will benefit from the reduced applications offered by Bt.[43] Despite the chronic periodicity in rainfall and the high fluctuations in year-to-year infestations, Bt could still make sense for this demographic of farmers over the long-term: functioning as affordable insurance against spikes in pest outbreaks, where increased costs in one year are absorbed by increased benefits in another. But rates of pesticide applications are much lower for smallholder white maize growers: as low as 3 per cent in some sites in the same survey reported above. One recent survey of smallholder growers spread out among six districts in the Eastern Cape and KwaZulu-Natal reveals that many of these growers did not identify stem borer as a "serious problem."[44] When outbreaks do occur, poorer farmers who cannot afford pesticides often opt for cheaper, more available substitutes, such as using ash or washing detergents to deter insect pests.[45] For these farmers, the cost savings of reduced pesticide applications offered by Bt will be minimal, while the risk presented by the increased technology fee will be magnified.

Another important constraint for smallholder farmers is a lack of information around proper cultivation practices for GM maize. Case studies from the provinces of Eastern Cape and Limpopo suggest that dissemination agents lack the key information and capacity to properly inform farmers about planting practice, leading to uneven application of proper cultivation practices and widespread confusion regarding the differences between GM varieties and hybrids.[46] Another study revealed that farmers across the main growing areas in the Highveld continued to apply insecticides to their Bt maize crops (incidence rates were as low as 5 per cent and as high as 93 per cent, contravening recommendations to avoid this practice, as it is known to accelerate resistance among borers).[47] Most of the technical information regarding proper handling and growing protocols required by GM maize is provided via a user guide that is distributed to farmers, despite the fact that many smallholders are semiliterate or illiterate. One recent study reported that 71 per cent of farmers lacked the capacity to understand these written instructions on proper cultivation practices.[48]

A closer examination of the reasons farmers offer when they do choose to plant GM varieties underlines this gap in both the accuracy and availability of information. When asked to identify the major reason for choosing to plant GM maize, 80 per cent of farmers offered that it matured faster than local varieties, which was widely appreciated, as it allowed farmers additional flexibility to delay planting until after the arrival of the rains. But a short maturation period is a trait that had been bred into hybrids, making it a characteristic of the host germplasm, rather than of the GM event itself. The particular benefits offered by genetic modification – higher yields and superior insect resistance – were mentioned by only 29 per cent and 5 per cent of respondents, respectively.[49] Another study corroborated these findings, reporting that the most common benefits that farmers associate with genetic modification – including drought tolerance, fast growing and improved taste – were in fact common among GM and non-GM isolines.[50] This research concludes that most adopters were swayed by corollary benefits such as prestige, participation in seminars and wide seed availability from project promoters, and that these "pseudo-adopters" will turn away from this technology once these associated benefits cease. This disconnect between the actual traits offered by the GM variety and the perceived benefits observed by farmers on the ground resonates with anthropologist Glenn Stone's important assessment of how GM technology can result in farmer deskilling – that is, new breeding technologies can serve to undermine the valuable knowledge and skills that farmers use to shape on-farm decision-making.[51]

Even more worrying are a slew of recent studies suggesting that GM and non-GM varieties are interacting and interbreeding through a variety of uncontrolled channels. The first pathway stems from the close physical contact between GM and non-GM maize. In order to preserve stem borer resistance in Bt maize, farmers must plant a refuge of non-Bt maize alongside their Bt crop, which serves as the feeding ground for stem borers. Mandated requirements for planting refuges stipulate either a 20 per cent refuge planted to conventional maize (which can be sprayed with insecticide), or a 5 per cent refuge area that is no-spray.[52] This arrangement helps to limit the development of resistance within the insect population, by ensuring that those rare, Bt-resistant individuals that survive on the Bt crop mate with susceptible individuals, thus reducing the possibility that two Bt-resistant individuals could mate and produce super-resistant offspring and perpetuate resistance in the larger population. Seed companies and local seed retailers communicate this imperative around proper practice. But a number of studies have emerged suggesting that compliance with these regulations is relatively low. One survey of the Vaalharts Irrigation Scheme in the Northern Cape Province found that fewer than 10 per cent of farmers planted refuges during the initial planting year of 1998. While compliance rates did increase over time, on average fewer than 30 per cent of farmers complied with refuge requirements in the first six growing seasons.[53] When refuges were planted they were overwhelmingly planted incorrectly: in 99.8 per cent of cases farmers allowed no spatial separation between the Bt field and the mandated buffer, which can facilitate minimal exposure of the Bt toxin to larvae that are able to migrate between the two sections, exacerbating resistance.[54]

Note that these data reflect practices amongst relatively large farmers on a formal irrigation scheme, most of whom were literate and had regular access to farmers' days organized by seed companies and extension agents. Compliance is much more difficult for smallholders, whose small plots are most often densely packed, one next to the other. Monsanto's user guide recommends a minimum spatial isolation of 400 metres between GM and non-GM varieties, while an independent assessment concluded that a minimum of 145 metres was needed to ensure a minimum threshold of gene flow.[55] Not surprisingly, compliance with these recommendations is low among the 77 per cent of South African growers whose average holdings are below two hectares:[56] one recent survey of seventy-eight farmers scattered across six growing areas scattered in Eastern Cape and KwaZulu-Natal found that none of the interviewed farmers had any knowledge of proper refuge strategies.[57] Without these spatial safeguards in place, high levels of pollen flow between the GM and non-GM varieties are impossible to prevent.

A second pathway that is accelerating rates of genetic mixing revolves around the prevalence of seed replanting, sharing, and recycling. Two recent studies quantify the pervasiveness of this practice, estimating it at 80 per cent and 76 per cent, respectively (though the authors of this second study suggest these are severe underestimates, and that this practice was common among *every single farmer* they encountered in their study).[58] They suggest that the prevalence of seed recycling practices were partially due to ignorance; some farmers for example, thought the restriction on recycling extended to all seed purchased from agricultural shops, even OPVs that had no restrictions attached to them. But other farmers seem to willfully disobey the limitation on recycling seed and resent the limitation on this long-standing practice. In the Eastern Cape, farmers refer to both GM and hybrid maize as *udlambhuqe*, a term that translates as "the maize that you eat until it is finished" – that is, seeds are not saved for replanting next year – suggesting that farmers understand these regulations but are choosing to ignore them.[59]

The third major pathway for mixing seed happens before the seed arrives on the farm. One of Fischer's recent papers suggest that seed retailers often have "nonspecific, unclear and occasionally incorrect" information on the seed they stock and distribute to farmers. Interviews with retailers suggest they had limited knowledge of the seed they were selling, and lacked basic understandings of proper management practices associated with GM seed, including proper refugia and the prohibition on seed sharing.[60] This confusion extended beyond a lack of proper information: one common practice among retailers is to repack GM maize into smaller bags that are more affordable for smallholders, leading to high rates of seed mixing and incorrect labelling.

The implications of this systemic confusion around and direct contraventions of proper management practice are significant. Each of these pathways accelerates the flow of transgenic material from GM varieties to their non-GM cousins, leading to higher-than-normal levels of gene flow.[61] While some gene flow from GM to non-GM varieties is inevitable in an OPV such as maize, the research surveyed above suggests that transgenic contamination is much higher due to the realities of the informal maize seed sector in South Africa. These high rates of germplasm mixing have led to spikes in development of resistance to these traits, as more widespread exposure to low levels of the Bt toxin has accelerated the genetically based decrease in susceptibility among the pest population.[62] To date, two separate instances of widespread resistance have been reported. The first generation of Monsanto's Bt maize with the Cry1Ab gene was pulled from the market only eight years after its initial release, with levels of resistance measured at between 43 per cent and 64 per cent at one site.[63] A separate incident was reported in subsequent years, when damage from stem borers on Bt maize jumped from 2.5 to 58.8 per cent in only two growing seasons, at which point more than half of those farmers planting Bt maize decided to spray their crop with insecticides.[64]

Further compounding these accelerating rates of resistance is the low starting point for toxicity at the time of commercialization, which hovered between 97 per cent and 98 per cent at the time of introduction.[65] As a comparison, the United States Environmental Protection Agency indicates that in order to satisfy the criteria for high-dose toxicity, Bt plants should kill 99.99 per cent of susceptible insects in the field. High-dose toxicity is an essential part of the overall strategy for minimizing the development of resistance: the dose of toxin ingested by insects needs to be high in order to kill all of the hybrid progeny that result from the mating of the rare resistant individuals with the abundant susceptible ones from the nearby refuge. This strategy only works if the initial dose of toxin ingested by the resisted individual is large enough that it kills the resultant progeny, which is why the approach is known as the "high-dose refuge strategy" and why the minimum dose mandated is so high (the assumption being if the initial dose is not high enough to kill 99.99 per cent of all susceptible individuals then it will be insufficient to kill nearly all hybrid progeny).[66] What is more, since the rate of survival for the hybrid progeny is higher than for their susceptible counterparts, the inheritance of resistance becomes a dominant, rather than a recessive, trait, which further exacerbates its pervasiveness throughout the population as a whole.[67]

Once resistance is entrenched in the pest population, there are two avenues for delaying its spread and containing its reach. The first involves increasing both the size and abundance of refuges – not an option in South Africa where compliance with the more modest initial refuge regulations was incredibly low. The second strategy to counter escalating resistance is more transgenesis. In 2013, Monsanto released a new version of Bt maize that combines two toxin-producing transgenes – Cry1A.105 and Cry2B – to replace the now-ineffectual single transgene, a practice known as gene pyramiding.[68] But while this new variety's dual trait structure will certainly provide a short-term boost to preventing resistance, it remains unlikely that it will be sufficient to counter the issues of low dose and farmer practice that are at the root of this problem, especially given that these new pyramids will be grown concurrently with the old single toxin plants, which diminishes their ability to delay resistance.[69] Other researchers have argued that this reliance on new generations of genetic modification to overcome the problem of resistance is both temporary and expensive for farmers, especially the poorest and most vulnerable who cannot afford the increased fees associated with these more complex technologies.

Inevitably, farmers get caught on the "transgenic treadmill," in which they are compelled to buy new and improved G M versions in order to maintain existing production levels.[70]

Another important issue in the debate around the potential for G M maize to help South African farmers is the variety into which this technology was inserted. During its first decade of cultivation, Monsanto's Bt construct was inserted into two American hybrids, first C R N 4559 and then C R N 3505. Both did extremely well in North America: these are high-yielders, with medium growth periods and very good drought tolerance. But their suitability in South African growing conditions was unknown when they were introduced: when C R N 4559 was first introduced in 2001/02, researchers tried in vain to obtain its nearest isoline (C R N 3549) to use as a counterfactual in trials, but were unable to access it because this variety was not planted at all in South Africa. It is thus not surprising that these newly introduced varieties underperformed relative to varieties that had been bred specifically to respond to the particularities of the South African climate. For instance, farm-level studies undertaken in the eastern province of KwaZulu-Natal reveal that during a period of particularly low rainfall, yields from Bt maize were outmatched by the most popular conventional hybrid throughout much of KwaZulu-Natal, P A N 6043, sold by Pannar.[71] This particular conventional variety was favoured by farmers largely due to its high levels of adaptability to drought and stress tolerance (so much so that it was used as the prototypical "stress tolerant variety" in a series of water stress experiments). In times of high stress, farmers largely turned away from Bt towards locally adapted hybrids and O P V s that offered better tolerance to the slew of abiotic and biotic stressors that afflicted maize production in their particular growing area.

Over the past few years all three of the G M traits (Bt, H T/Roundup Ready, and B R/stacked) have been incorporated into locally used hybrids, which should help to avoid the pitfalls associated with foreign-bred host varieties. But, as Fischer and Hajdu demonstrate, the majority of smallholder farmers continue to favour the lower-yielding but more resilient O P V s, which are cheaper, can be saved year to year, and tend to be adapted to less-than-optimal growing conditions.[72] Their research suggests that one of the greatest limitations of hybrids (either locally adapted or not) is their inferior storage capacity. They found that 72 per cent of farmers considered G M maize to be more susceptible to grain weevils, a common limitation associated with modern maize hybrids that lack the hard grain of OPVs that are difficult for weevils (the major pest of post-harvest losses) to penetrate.

Despite these numerous advantages, a GM version of an OPV has never been created. To date, there has only been one attempt to develop a GM OPV: the Insect Resistant Maize for Africa (IRMA) project (examined in more detail in chapter 5). This joint venture was started in 1999 as a collaboration between the Kenya Agricultural Research Institute (KARI), the International Maize and Wheat Improvement Center (CIMMYT), and the Syngenta Foundation. It sought to use both conventional breeding and genetic modification to create insect-resistant maize varieties for Kenyan farmers. What is unique about this case is that it remains the only project that has attempted to use publicly derived Bt genes (so there are no issues around technology fees or prohibitions on replanting or recycling), as well as the only example in which OPV host varieties were attempted alongside hybrids. The IRMA project completed three full phases and produced a number of viable insect-resistant varieties via conventional breeding. But in the end scientists were unable to produce a commercially viable GM OPV. According to one of genetic modification's fiercest critics, IRMA was discontinued because of both technical reasons (those Bt genes available in the public domain proved insufficient barriers to borer damage) and agronomic ones (reusing Bt seed would accelerate insect resistance, undermining the technology's viability within a couple of years).[73] Reports produced by the IRMA research team corroborate this version of events, suggesting that the project was successful in finding Bt events that were resistant to four of the five major borer pests in Africa, but then stalled after they were unable to find an event in the public domain that showed significant resistance to the African stem borer.[74] In an interview, the project's lead scientist suggested that the difficulty in managing the efficacy of a GM OPV – which is less stable and more difficult to segregate than a hybrid – was a bigger hindrance than the issue of intellectual property.[75] Thus, the IRMA case underscores the challenges of integrating GM technology into OPVs, which, by their nature, violate the intellectual property and the ecological integrity of genetic modification by allowing replanting and recycling of seed.

Taken together, the issues surveyed in this section underline the incongruences between the genetic, economic, and legal requirements for GM technology and the social and ecological realities that shape smallholder production in South Africa. All genetic modification cultivation in South Africa is regulated by the Genetically Modified Organisms Act of 1997, which stipulates that any farmer choosing to grow GM seeds must sign a technology licensing agreement, in which they certify their compliance with the conditions granted to the permit holder (generally the seed provider), in order to mitigate negative health or environmental impacts. Such restrictions include strict adherence to an insect-resistance management program that generally involves the planting of buffer areas around GM crops; a ban on seed saving, exchanging, or replanting; and generally assuming liability if these conditions are violated. But there are a number of characteristics of smallholder maize farming in South Africa that make adherence to such regulations unlikely. The small farm sizes, close proximity of neighbouring farms, and inability or unwillingness to maintain refuges make trait segregation unrealistic. Practices of sharing and recycling seed are ubiquitous. Mixing is inevitable. These quotidian practices undermine the integrity of the GM trait. For genetic modification to be a viable solution over the long-term, it must be contained and regulated within the formal seed sector, which is incompatible with the current realities of smallholder practice in South Africa. There exists a fundamental incompatibility between the technology and the farming system in which it is supposed to succeed.

BT MAIZE IN EGYPT

Egypt remains the only country other than South Africa to have commercialized GM maize on the continent. Maize is an important staple in Egypt, where it is planted on more than 750,000 hectares of land, producing in excess of six million tons annually. Over 90 per cent of all maize planted is devoted to white maize, most of which is consumed locally. This heavy imbalance leaves a sizeable gap in yellow maize, which is needed as feedstock: annual production deficits are so severe that imports exceed five million tons annually, at a value of over US$1.6 billion.[76]

The interest in the potential for Bt maize in Egypt was due largely to advancements by and interest from domestic scientists and regulators, who initiated a partnership with Pioneer Hi-Bred, facilitated by the Agricultural Biotechnology Support Project (ABSP) and USAID-Cairo, with the aim of commercializing novel Bt proteins and genes created by Egyptian scientists.

The first and only G M crop to be commercialized in Egypt was Ajeeb YieldGard (Ajeeb Y G), a yellow hybrid maize that emerged by crossbreeding Monsanto's Bt yellow maize hybrid (M O N 810) with the local maize variety Ajeeb. Ajeeb Y G was the result of a partnership between Monsanto and the Cairo-based seed company Fine Seeds International. Fine Seeds International submitted an application for the regulatory approval of this G M variety in 2000. A formal biosafety assessment was initiated in 2004 by the Egyptian N B C, though it is important to note that no biosafety legislation existed at that time: instead, this process was governed by the Cartagena Protocol and the European guidelines for risk assessment (which were imposed unilaterally by then president Hosni Mubarak's government).[77]

Field trials with Ajeeb Y G were initiated in 2002. Initial results from four separate experimental stations were promising: Ajeeb Y G offered nearly 100 per cent resistance against three species of maize borers; one report suggested that yields were 30 to 40 per cent higher than conventional varieties.[78] The ministry of agriculture granted final approval for this new variety of G M yellow maize in 2008, making Egypt the second African country after South Africa to commercialize G M crops.[79] Seven hundred hectares were planted in the inaugural year of 2008, which ballooned to 2,000 hectares in 2010. Initial reports were favourable, reporting a per hectare benefit of US\$281.[80] Anecdotal evidence among smallholder growers compiled by I S A A A confirmed that these benefits accrued even to those with only minimal holdings, reporting increases in yields of 25 per cent and substantial reductions of pesticide use.[81] Later assessments suggest that these field plantings offered yields between 50 per cent and 150 per cent higher than conventional hybrid corn.[82]

But this promising trajectory quickly turned sour. Within three years of commercialization there was increasing public resentment against the project, with only a muted defence from promoters.[83] Accusations began emerging of massive shipments of G M maize both into and out of the country without proper government approval or oversight. In January 2012, the ministry of environment seized and destroyed a shipment of forty tons of GM seed imported from South Africa, arguing that the approvals process it had undergone – the shipment was originally checked by two advisors from the ministry of agriculture, both of whom had previously served as presidents of the Agricultural Research Center, which oversees A G E R I – was improper and contravened both the constitution and the Cartagena Protocol.[84]

Two months later, on 8 March 2012, the minister of agriculture issued Decree no. 378, which suspended both the planting and importation of all GM maize until both the ministry of health and the ministry of environment could complete a full risk assessment.[85] The commercial release of GM maize has been on hold ever since.

How did the tide turn so quickly against GM maize in Egypt? According to the USDA Foreign Agricultural Service, this abrupt shift in policy was due to an "aggressive media campaign" that sought to demonize genetic modification within the public sphere.[86] The spike in anti-GMO sentiment is certainly accurate: over the past number of years public sentiment in the country had become increasingly vocal against Monsanto's presence in the Egyptian marketplace. Some of this opposition was fuelled by negative coverage around health concerns: in a flurry of studies released in 2012 and 2013, Ajeeb YG was fed to rats, revealing a plethora of negative health implications, including liver damage, congestion of the spleen, and enlargement of the small intestine.[87] These concerns were picked up quickly in the Egyptian media: the USDA Foreign Agricultural Service claims that the 2012 ban stemmed directly from a three-day television special that aired in March 2012, alleging that Ajeeb YB yellow corn caused cancer and liver damage.[88]

Other opposition targeted Monsanto itself. A March Against Monsanto on 25 May 2013 drew over one hundred protesters. Coordinated by a local NGO, Bozoor Balady (Seeds of My Country), the centrepiece was a massive dragon with a cob of corn in its mouth, symbolizing the global power of the multinational corporation and its current efforts to eat all of the world's corn supply. Protesters chanted: "The crops of our country are the solution, Monsanto means humiliation."[89] The following morning, a popular Egyptian talk show, Zay al Shams, hosted a thirty-minute program dedicated to the issue of GM seeds. Guests included members of Greenpeace and farmer representatives, who emphasized the lack of transparency around the technology's penetration into Egyptian markets while stressing the controversial claims over the link with cancer.[90]

But the USDA's assessment of Egypt's abrupt turn with Bt maize overestimates the role of opposition forces and underestimates the role of broader political forces. This abrupt shift in policy and public sentiment towards GM maize reflected the broader political changes that were taking place within the country at that time. In February

2011, the anti-government fervour that had begun two months earlier in Tunisia led to the toppling of Hosni Mubarak's autocratic regime. Reports suggest that this landmark regime change played a pivotal role in undermining support for GM maize: many prominent private businesses – including Fine Seeds International – were viewed with hostility and suspicion after the uprising, accused of cultivating close connections with the Mubarak regime. The relationship between the new transitional government and this holdover from the previous era suffered due to increasing mistrust: both sides were increasingly skeptical of the other's integrity and intentions.[91]

The collaboration between Monsanto and Fine Seeds International was perceived as a holdover from the previous era, an outdated venture that was aligned with the old regime and was tainted by its legacy. Interviews with those familiar with this public-private venture reveal that even some employees of Fine Seeds International shared this view: one representative expressed his frustration with these power imbalances in vivid terms, referring to Monsanto as "our masters," a continuation of the "old before-the-revolution mentality" in which Monsanto representatives emulated Mubarak's autocratic rule and treated their Egyptian partners as "slaves."[92] Another employee of Fine Seeds International echoed this sentiment, arguing that Monsanto was never truly committed to the Egyptian market.[93]

The Egyptian case underlines the inevitable intertwining of science and politics in the debate over GM maize in Africa. Proponents of the technology argue that the promising early trajectory of GM maize was derailed by unfair and untruthful critiques levied by a select group of protesters. In order to reverse the ban, a "solid and sustained communications drive by scientists and stakeholders" is needed, in order to "build public understanding of agricultural biotechnology, public confidence, and a political climate that would allow the resumption of product development and commercialization of GE [genetically engineered] crops."[94] But it is the third element here – the changing political climate – that was most influential in shaping this trajectory. Fairly or unfairly, genetic modification was perceived as an alliance of the old guard, and was explicitly rejected because this venture was tainted by its affiliation with the newly deposed president. This case thus offers important lessons for the debate around genetic modification in Africa more broadly: the success or failure of a GM variety depends on the political context in which it emerges.

CONCLUSION: MOVING BEYOND YIELDS

Despite the stalled progress in Egypt and the spotty success among smallholder farmers in South Africa, many boosters remain confident in the potential for GM maize to help reduce poverty and hunger across Africa. Much of this optimism is focused on the possibilities of extending GM maize into other countries in eastern and southern Africa, where maize is the primary carbohydrate staple for the vast majority of the population. An economic surplus model based on data collected in Kenya, for instance, predicts profits in excess of US$200 million over twenty-five years once GM maize is adopted.[95] Another team of researchers using a different economic model forecasts that while overall yield increases to smallholders in the major maize-producing nations of southern Africa will be low, they will be sufficient to address current gaps in domestic food supply and thus enhance overall food security in the region.[96]

The empirical evidence from South Africa and Egypt suggests that these predictions are overly optimistic. While the data showing the potential economic benefits to large-scale adopters of yellow maize is convincing, the stalled adoption rates and increasing questions about the impact of this technology on small-scale producers shed doubt on the potential for this technology to help those farmers who need it most. This chapter has surveyed some of the major constraints to GM maize and smallholder farming, including a lack of proper information and understanding regarding best practices and the pressing issue of transgenic resistance. Most crucially, the economic and ecological integrity of the technology is incompatible with current practices of poor growers, whose small holdings, close proximity to neighbouring farmers, and predilection for sharing and recycling seed clash with the trait segregation practices that are foundational to the technology's success.[97] An emphasis on political ecology reminds us that it is these interactions between biophysical and social systems that ultimately determine agricultural development outcomes.

The breeding programs that offer the most promise for smallholders are those that are able to break the cycle of the narrow focus on yields that has shaped experimental efforts since colonial times. One of the pioneering efforts in this regard is the Southern African Drought and Low Soil Fertility (SADLF) project, which used the low-input and high-stress environment facing smallholder farmers as the starting point for its campaign. The focus of these efforts was on OPVs that were the progeny of dozens of varieties drawn from the CIMMYT genetic pool, rather than the result of recurrent selection of a single variety (in the case of hybrids). Promising varieties were then assessed within conditions of low moisture and low soil fertility in order to bolster the hardiness needed by smallholder growers. The result, recounts historian James McCann, was the emergence of a "flinty" variety that reflected greater genetic diversity and resembled the hardy landraces that had dominated throughout the sixteenth to nineteenth centuries, a variety that proved both harder and hardier than the most popular hybrids.[98] These new varieties had characteristics that appealed to smallholder growers, including better storability, increased malleability that made them easier to pound, a favourable taste, and minimal loss during the milling process. These OPVs also proved to be more resilient under conditions of stress, such as low fertilizer content and water shortage.[99] These new varieties represented a monumental shift in the process of maize breeding: breeders inversed the prioritization that has produced generations of high-performing hybrids by giving precedence to the real-world conditions facing African farmers.

These efforts persist today within the Drought Tolerant Maize for Africa (DTMA) project, which continues to focus varietal improvement around the traits that matter most to smallholder farmers, including yield stability, drought resistance, and low soil nitrogen tolerance.[100] Breeders are continually modifying experimental programs to integrate farmer preferences, including improved storability, early maturation, and resistance to major maize diseases. Farmers also have the option of recycling seed without the heavy yield losses associated with replanting hybrids.[101]

The SADLF and DTMA projects are committed to the production of resilient varieties that can thrive within the real-world conditions that confront small-scale farmers. In contrast, existing varieties of GM maize remain steeped in the long tradition of elite breeding, whereby technologies are designed to thrive in agricultural systems that fail to resonate with smallholder realities. GM maize makes sense for large-scale producers with access to capital, and who engage in mechanized farming with high-precision irrigation. But this is not the reality for most smallholders in Africa. In order to make a difference for these farmers, GM maize will need to be able to respond to the array of stressors – both ecological and economic – that characterize their agricultural systems.

SECTION THREE

GMO 2.0:
Second-Generation GM Crops Targeting Traits and Crops That Matter to Poor Farmers

5

Two Crucial Precedents:
Virus-Resistant Sweet Potato
and Insect-Resistant Maize in Kenya

This chapter chronicles the first two attempts to create GM versions of African staple crops: the GM sweet potato and the Insect Resistant Maize for Africa (IRMA) project. Both projects were initiated in Kenya in the 1990s, and both proved to be important harbingers for the more extensive and expensive wave of GM staple crops that are scheduled to be released soon. Each of these precedents reveal important lessons that foreshadow the much larger investment in GM orphan crops that accompanied the push towards Africa's Gene Revolution. This chapter outlines each of these initiatives in turn, unearthing the lessons learned from these first attempts to use genetic modification as a means of improving the array of staple crops that are most vital to African smallholder farmers.

THE VISION: P3S AND AGRICULTURAL INNOVATION IN AFRICA

The current model of technological diffusion revolves predominantly around public-private partnerships, also known as P3s, which have emerged as the preferred model for making GM versions of staple crops available to African smallholder farmers. The most often cited rationale behind the rise of P3s is the stagnating level of public investment in African agriculture, which has created deficiencies in expertise, infrastructure, and capacity.[1] The result is a highly imbalanced playing field: most of the world's novel, cutting-edge technologies are owned by the private sector, while the public sector's ability to bring new innovations to market has been gutted by decades of underfunding.[2]

P3S originated as a strategy for overcoming these increasingly pronounced inequities between the private and public sectors. The private sector partner commits to providing fundamental scientific data, often in the form of patented technologies, as well as relevant expertise, while the public sector partner takes responsibility for the direction and management of the experimental process. In return, private sector participants can boast about how they are much more than profit-making enterprises, and reap the public relations boost that comes with supporting such humanitarian endeavours. In the words of a former Monsanto executive (and current executive at the BMGF), "the private sector's motivation and commitment to share is based on a variety of factors, including philanthropic interests, humanitarian concerns, employee initiative, good public relations, and new business opportunities."[3] A further incentive for the private sector is that this commitment to P3S increases the pressure to create regulatory regimes that are congruent with international standards in terms of intellectual property and stewardship, which are both vital to protecting the industry's investments in innovation.[4] Critics argue that P3S serve to enlist public sector partners as allies in the private sector's bid to create permissive regulatory structures that will accelerate expansion into new markets and lead to larger profits in the long-term.

For public sector partners the major reward comes in the form of access to proprietary technology they would otherwise not be able to obtain and capacity building for their own burgeoning science and technology sectors. Over the long-term, proponents argue that the technology transfer, capacity building, and knowledge sharing built into P3S will encourage public sector interventions to address market failures.[5] Another highly touted benefit of P3S for the public sector is that they mobilize private sector resources towards crops that have long been ignored by both innovation and investment. P3S are trumpeted as a means of stimulating innovation in so-called orphan crops, staple crops that provide nourishment to hundreds of millions of Africans yet have been virtually abandoned by private sector investment, mostly because there is no profit incentive when crops are eaten almost exclusively by poor people. P3S are advanced as a means of addressing these "market failures ... wherein the social benefits of research exceed the private benefits."[6] Creating P3S that use GM technology to improve productivity in these long-ignored crops has emerged as a central pillar of Africa's Green Revolution.[7]

This renewed emphasis on a partnership approach has also received its share of criticism. The most common reproach relates to asymmetrical power relations, which threaten the genuine nature of these partnerships. Critics contend that the more powerful actors – invariably the private corporations and development donors – end up setting the terms and agenda to favour their own interests by prioritizing issues such as free trade, regulatory harmonization, and access to new markets. The least powerful actors – the farmers themselves, who are positioned as the intended beneficiaries of these partnerships – are shut out of negotiations around key issues such as design, targets, outcomes, or evaluation.[8] Left-leaning NGOs denounce P3s as yet another vehicle that enhances benefits for the rich and powerful, while ensuring that any risk is borne primarily by the poor and vulnerable.[9] When such ventures do eventually reach out to farmers, they disproportionately benefit commercially oriented farmers or farmer organizations, who tend to have larger holdings, be more heavily capitalized and better positioned to meet quality control requirements. As such, P3s fall into the trap of missing the vast majority of African smallholders, 90 per cent of whom do not fulfill these minimum criteria.[10]

VIRUS-RESISTANT SWEET POTATO AND INSECT-RESISTANT MAIZE

Initiated in 1991, the GM sweet potato project was designed to use GM technology as a means of mitigating damage due to disease, which some proponents claimed was as high as 80 per cent for smallholder farmers in Kenya.[11] The project was designed to create a transgenic sweet potato that was resistant to the sweet potato feathery mottle virus (SPFMV), one of the three major viruses affecting the crop in East Africa. The experimental work was undertaken as a partnership between KARI and Monsanto, which donated their viral coat protein gene that confers viral resistance when transferred and expressed in a plant genome. Financial and institutional support was provided by USAID and ABSP.

The partnership's origins remain a source of contestation: Florence Wambugu, the project head, insists that KARI turned to genetic modification after concluding that biotechnology was the only viable solution to combatting damage due to disease, while the anti-GMO organization GRAIN traces the origins of the project back to Monsanto and USAID, who were eager to produce a GM crop that addressed issues of hunger and poverty in Africa.[12] An interview undertaken with a source close to the project suggests that it was driven by supply more than demand: KARI's director at that time was "very pro-biotechnology"; he paired up with investors on the side and set up his own private tissue culture business. At the same time, "Monsanto and other [seed companies] were looking for entry into Africa."[13] USAID also played an important role as facilitator, working actively to create entry points for American seed companies. According to Odame, Kameri-Mbote, and Wafula, Monsanto itself provided US$2 million for the project, which at the time was the largest single biotechnology research contribution by the private sector.[14]

Hope was high for this new partnership. Matin Qaim, a prominent agricultural economist who specializes in measuring the potential yield benefits associated with GM crops, undertook an ex ante study that found that GM sweet potato would allow farmers to increase their earnings by 28 per cent, with a total annual benefit of US$5.4 million accruing to smallholder farmers.[15] Others were less restrained. The project's leading scientist predicted that controlling SPFMV could lead to a doubling of yields and an additional 1.8 million tons of production, amounting to over US$500 million in annual revenue.[16] A feature in *Forbes* magazine in 2002 reported that the preliminary results obtained in field trials were "astonishing," proclaiming that "on a continent where population growth outstrips food supply growth by 1 per cent a year, Wambugu's modified sweet potato offers tangible hope."[17]

The first research phase dedicated to the collection and selection of germplasm and genetic transformation took place at Monsanto's laboratories in St. Louis, Missouri, from 1991 to 1997. The lead scientist on the project was Wambugu, who had been serving as the coordinator of plant biotechnology research at KARI before beginning a position as a postdoctoral research associate at Monsanto from 1991 to 1994. Wambugu continued leading this consortium until 2001, while serving concurrently as Africa regional director of the ISAAA from 1994 to 2001. Using six Kenyan sweet potato varieties as a starting point, the experiments focused on introgressing the virus coat protein donated royalty-free from Monsanto. By 1997, 195 lines of CPT 560 were transformed. These initial tests in the US demonstrated that the GM varieties had "good yield potential with yield increases of 18 per cent."[18] Wambugu shared these promising early results via op-eds in *Nature* and the *New York Times*, followed by appearances on CNN and CBS's *60 Minutes*.[19]

Importation and approval did not go as smoothly. Despite being ready in 1997, the transgenic lines did not actually arrive in Kenya until 2000, due to a lengthy review process undertaken by Kenya's NBC, which suffered from constrained capacity. KARI imported CPT 560 in the year 2000, due in large part to considerable pressure applied by Wambugu's influential position with the ISAAA.[20] The new GM versions were then integrated into CFTs at experimental stations across the country. The intent was to evaluate their performance as compared with popular traditional varieties.

Field results were disappointing. The first two years confirmed that the transformation process was successful, but the gene demonstrated low levels of resistance, prompting Wambugu to speculate that a homegrown Kenyan virus would be needed for transformation, rather than relying on the one supplied by Monsanto. Results from the experimental trials also showed that the traditional sweet potato varieties produced significantly more tuber than their transgenic counterparts, further undermining the potential of these GM varieties to help poor farmers.[21] In 2004, news of the failure became public and experiments were abandoned shortly after, with over US$6 million having been invested over the lifetime of the project.[22]

The second P3 initiated to create a GM version of an African carbohydrate staple crop was the IRMA program. Initiated by CIMMYT – the CGIAR responsible for maize breeding – IRMA was conceived of as a continental-wide solution to the problem of stem borers, a collective of pests that together account for upwards of 15 per cent of total yield losses.[23] The project objective was to make existing licence-free Bt maize technologies accessible and affordable to smallholder farmers, which, according to researchers, would translate into yield increases of more than 40 per cent.[24] The project was launched in 1999, guided by CIMMYT in collaboration with KARI. Financial support came from the Novartis Foundation, which was subsequently consolidated into the Syngenta Foundation for Sustainable Agriculture (SFSA) following a series of mergers of acquisitions.

SFSA provided just under US$10 million in funding to support IRMA in its first two phases of operations.[25] Phase one (1999 to 2004) focused on capacity building and proof of concept. First, IRMA made considerable investments in infrastructure and training to build the local competencies needed to undertake the experimentation life cycle with Bt maize. IRMA funded the building of a biosafety level two laboratory, a biosafety level two greenhouse complex, and an open quarantine site, and supported the training of five molecular biologists.[26]

During this first phase, IRMA breeders were successful in developing source lines of the key Cry1Ab and Cry1Ba genes, which were obtained from the University of Ottawa.

IRMA's phase two (2004 to 2008) was designed to deliver the first set of improved varieties to smallholder farmers. But it was hampered by a series of setbacks. First there were unhappy rumblings from the donor, the SFSA, whose president began voicing complaints about poor record keeping and funds being directed towards product development that were unrelated to the project's core aims.[27] This led the SFSA to reduce core funding for phase two, requiring IRMA to solicit other donors to make up this shortfall.[28] Invitations to other potential donors were unsuccessful, except for US$440,000 from the Rockefeller Foundation pledged to help the project achieve its phase two goals. But even with this supplement the project had a shortfall of just under one million dollars, which forced them to scale back activities that were deemed tangential to the project's core goals. Still, IRMA breeders were able to proceed with initial greenhouse trials and the first round of field trials was planted in 2005. The IRMA executive committee visited the field trials two weeks after they were planted, and were surprised to find extensive gaps between rows, which the technicians explained were due to damage by grubs. According to first-hand accounts, the executive committee began brainstorming over possible countermeasures while they were on site, but left before making any final recommendations. Following their departure the chief technician decided to act on the "casual statements" made by executive committee members and took it upon himself to apply a broad-based insecticide to inhibit grub damage on both the Bt and the control plans, undermining the validity of the trials.[29] Once this mistake became public, Kenya's National Biosafety Committee recommended terminating and subsequently restarting the trials. This mini-scandal precipitated all sorts of public reaction, including pointed critiques from a high-ranking official in the Kenyan ministry of agriculture who accused the project of side-stepping government regulations, paying insufficient attention to long-term impact data, and rushing experiments to showcase project success. A member of parliament who was a long-time opponent of genetically modified crops levied further accusations that KARI officials were taking bribes and releasing GM food into the public supply. The fallout seems to have shaken internal confidence within the project.[30]

These setbacks were further compounded by unexpected challenges related to intellectual property rights. CIMMYT obtained the Cry constructs from a biochemist at the University of Ottawa who had succeeded in optimizing the coding for the DNA sequences for three Cry proteins.[31] The inaugural negotiations that took place in 1996 set out a material transfer agreement (MTA), which stipulated that the proteins were to be used exclusively for research purposes. At the same time, the IRMA project head commissioned a proactive survey of potential intellectual property rights issues, which concluded that no patents for Bt maize had been filed in Kenya, leading him to conclude that "no patent restrictions are expected."[32] It was not until the end of phase one that IRMA officials recognized the potential obstacle presented by intellectual property rights: sometime in 2002 or 2003, the IRMA team first clued in to the fact that the events donated by the University of Ottawa were synthesized versions of the Cry constructs and thus in every other way "identical" to the proprietary technology owned by Monsanto.[33] CIMMYT lawyers were concerned that any potential release would not be covered by existing arrangements, and ultimately "advised against it."[34]

At this point project leaders commissioned a CIMMYT lawyer to approach the University of Ottawa to transfer the current MTA into a freedom to operate agreement, which would allow breeders to eventually commercialize IRMA maize using the most promising construct donated from the University of Ottawa, the Cry1Ab gene. The university didn't respond until over a year later, at which point it made it clear that it would only waive the MTA if the project "secured declarations of agreement from multinational seed companies Syngenta, Monsanto, Bayer and DuPont."[35] Upon learning this, IRMA sent a negotiator to Monsanto in 2005 to request the rights to use this licensed technology. Preliminary negotiations were encouraging, as Monsanto was willing to discuss a maximum threshold that would allow smallholder growers access to IRMA while ensuring that larger-scale growers would purchase their seeds from Monsanto directly. Another recommendation from Monsanto was to limit the extent of the licence-free agreement to a few years, the logic being that after this introductory period farmers' incomes would rise to a level where they could afford to pay for the licence. Ultimately, Monsanto denied this request, due primarily to concerns around liability and redress.[36]

The turmoil over intellectual property rights was a major challenge and disappointment for the IRMA team. As one CIMMYT researcher affiliated with the project acknowledged in 2006, "After six years we have no argument for continuing with the CryIAb events. The entire environment has changed. We have been forced to acknowledge that events from the public domain will not automatically be available if we want to do more than research. Therefore, we must find someone who will give us access to an effective event."[37] But eventually this intellectual property rights hurdle became irrelevant. While IRMA scientists were determined to find a way to use the licence-free CryIAb protein they had received from the University of Ottawa, repeated tests demonstrated little resistance to the target pest of stem borers. The IRMA team had originally imported Bt maize leaves grown in Mexico and conducted bioassays to measure their efficiency against the five major classes of Kenyan stem borers, with subsequent testing in greenhouses and field trials. Each subsequent test revealed similar results: considerable resistance to the four minor borer pests, but low resistance for the most damaging of the East African stem borers, *Busseola fusca*, which accounted for 82 per cent of total damage.[38] IRMA breeders attempted to stack CryIAb and CryIAc proteins to overcome this hurdle. By the end of phase one this strategy had produced better results (i.e., 82 per cent resistance to *Busseola fusca*) but was unable to get any closer to complete resistance. By 2003, IRMA breeders were resigned to failure: "none of the tested Bt genes were adequately effective against *Busseola fusca*."[39]

In response to this series of challenges, the IRMA project undertook radical alterations in 2006 designed to resuscitate the project. The technical advisory board explored the idea of using Monsanto's MON 810 gene, which was shown to be resistant to *Busseola fusca*. Monsanto confirmed that they would not oppose use of the gene; however, there was fundamental disagreement over terms – in particular, if they were to grant poor farmers a licence-free version, and charge commercial farmers for the licence, how would they decide who was poor enough to be a beneficiary, and how would they enforce this arrangement?[40] In addition, Monsanto was unwilling to negotiate any deal beyond the use of hybrids, which undermined the potential development of GM OPVS, which was a core element of the original project.[41]

The abandonment of OPVS in favour of hybrids created a split within the IRMA project. The representative from the SFSA was adamant that maintaining GM versions of OPVS was "almost impossible to achieve in practice" because smallholder planting strategies involved mixing, recycling, and replanting OPV seed, which would accelerate cross-pollination with non-Bt neighbours and lead to a gradual loss of resistance.[42] Other members of the IRMA team were less reluctant. One suggestion that emerged was to engage farmers in large-scale training programs designed to reduce the risk of cross-pollination by planting the GM OPVs in the centre of their fields and avoiding selection of plants with visible insect damage.[43] Another researcher reminded her colleagues that the issue of farmers replanting and recycling seed persisted with hybrids as well, "although everyone advises them against it."[44] In the end, though, Monsanto remained firm in its unwillingness to extend the project to cover OPVs.

IRMA's third and final phase (2009 to 2013) reflected this foundational shift, as the program abandoned its focus on publicly sourced GM events and its target of creating GM OPVs. Instead, phase three focused on the identification and commercial release of conventionally bred insect-resistant germplasm. This new emphasis produced significant results: thirteen stem borer-resistant varieties were released to farmers, and an additional ten pest-resistant hybrids were nominated for national trials.[45] IRMA's major benefactor, the SFSA, seemed very pleased with these outputs: one high-ranking official remarked, "A very valuable combination of 'conventional' traits now seems to be maturing. They can help maize cope better with drought, disease and insects. We want to see how we can take these traits forward in seed systems, getting the best available material into smallholders' hands."[46] An end-of-project assessment by CIMMYT calculated the annual project benefits at between $19 million and $388 million, with a benefit-cost ratio that could be as high as 94 (meaning that for each dollar IRMA spent, farmers would gain $95).[47]

LESSONS LEARNED

The P3s that led to the aborted GM sweet potato and IRMA maize offer some important lessons for the much more expansive wave of GM staple crops across Africa that came crashing down over a decade later. Let's begin with sweet potato. The first issue was the choice of trait; that is, the decision to focus on sweet potato feathery mottle virus (SPFMV), which along with sweet potato chlorotic stunt virus (SPCSV) constitute the co-infection known as sweet potato virus disease (SPVD). On its own, SPFMV does not cause extensive damage; for example, in a controlled experiment SPFMV alone did not reduce yields. When combined with SPCSV, however, yields were reduced by 50 per cent.[48] The mechanism of this co-infection was not fully understood in the early 1990s when the transformation of the sweet potato was first undertaken, and the choice of zeroing in on SPFMV as the lynchpin of the experimentation program seems like a reasonable strategy, even in hindsight. What is more surprising is that Wambugu continued to focus exclusively on the singular pathway through which SPFMV leads to SPVD, even after research had clearly identified the decisive role SPCSV played within the infection. In 2003, Wambugu summarized the progress of her research program in the peer-reviewed journal *Nutrition Reviews*, emphasizing that the "key constraints, however, are diseases, of which SPFMV and weevil are the most important. SPFMV does not cause major losses by itself, but is frequently found in combination with one of several other viruses, resulting in a synergistic increase in disease severity. On susceptible varieties, yield loss caused by the virus complex can be as high as 80 per cent."[49] This semantic shuffle between the devastation caused by the virus complex (SPVD) and the actual focus of the experiments themselves (resistance to SPFMV) lies at the core of the project's failure. Proponents like Wambugu continued to trumpet the potential of the SPFMV-resistant varieties years after other researchers had uncovered the "synergistic interaction" between SPFMV and SPCSV.[50] In other words, the researchers knew they had focused on the wrong virus – or, more accurately, they realized they had focused on a single virus rather than the complex that actually causes the damage – but continued to justify this approach even after other scientists had concluded that it was the interactions between the causal viruses that produced the most damage: "The gene construct combining both resistant components will be more likely to offer a practical solution for the control of SPVD in sweet potato."[51] The compartmentalized approach employed by Wambugu was still being lauded years later, despite the recognition that the breeding approach taken had failed to appreciate the complex dynamics that were at the root of this disease.

Why was this reductive logic so difficult to shed? The answer lies in a breeding approach that puts the solution before the problem. The starting point for the GM sweet potato project was the viral coat protein donated by Monsanto, an existing genetic construct with potential applicability in an orphan crop. In breeding parlance this is referred to as an "implementation technology," which required an "enabling technology" to transplant this gene, conferring virus resistance into the target crop. The GM sweet potato project was designed to be the enabler. This logic of transplantation underlay these first attempts at breeding for GM versions of African staple crops: "in this way, genes isolated originally for use in a major crop such as wheat or soybean can be readily modified for use in orphan crops."[52]

This same logic was evident in optimistic assessments of the potential benefits of using the Bt gene to reduce the impact of the sweet potato weevil. The problem is induced by the yield gap: seventy tons per hectare in the US, while average yields in East Africa are less than five tons per hectare. One of the major pests is weevil, which can reduce yields by 60 per cent. An implementation technology already exists: "Genes encoding insect-specific toxin proteins have been isolated from the bacterium *Bacillus thuringiensis* and transferred into several major and orphan crops (e.g., maize, cowpea) to confer resistance to certain insect pests. Researchers are now proposing to introduce these genes into sweet potato cultivars to confer resistance to the sweet potato weevil."[53] Economic models are then used to show how even a paltry investment of US$10 million in an enabling technology to apply Bt resistance to sweet potato weevil could yield staggering benefits: incomes could be increased by 38 to 40 per cent, and unit costs could be reduced by 20 per cent per hectare, generating gross annual benefits of US$121 million.[54] This same reductionist logic elevates the potential applications of existing proprietary technologies: quixotic predictions broadcast by outside experts and based on economic models that are largely divorced from farmers' realities.

The problem with this emphasis on outsider technology is that it underestimates – rather than builds upon – the various techniques that farmers use to mitigate SPVD. East African farmers have their own techniques for managing SPVD by planting resilient varieties that recover well from infection, such as New Kawogo, which account for more than a third of the total area under sweet potato in the heaviest growing regions of Uganda, as well as relying on cultural methods of control, including selecting for non-infected vines, a practice at which farmers have been shown to be particular adept.[55] Critics like Aaron deGrassi also emphasize the success of conventional breeding programs in producing high-yielding and disease-resistant varieties, but more recent surveys of the Ugandan National Sweet Potato Program, which ran from 1999 to 2010, reveal spotty results: all of these high-yielding varieties demonstrated only low to moderate levels of resistance and were deemed suitable only to a region with low SPVD pressure.[56]

This tendency towards putting the solution before the problem was further evident in the choice of viral strain used for the GM event. There are four distinct genetic strains of SPFMV: strain EA contains East African isolates but none from elsewhere, while the other three strains all contain a mix of isolates from multiple continents. The viral coat protein donated by Monsanto contained a strain known as RC, which was derived from an American strain. These insights regarding distinctions between strains were emerging in the scientific literature around the same time as the GM sweet potato program, and plant geneticists had long acknowledged the distinct traits of these different SPFMVs and the central importance of the genetic variability within sweet potato viruses.[57] Three research articles published by a single research team between 1998 and 2001 underline the agro-ecological incompatibility between different strains: they were able to prove that East African isolates of SPFMV demonstrate a "distinguishable phylogenetic lineage" that is able to overcome resistance demonstrated against the three other strains of SPFMV that occur elsewhere.[58] But breeders were locked into this choice; there was no resistance to the EA strain on offer, so they chose to persist with the resistance to the American strain in the hopes that its application could yield similar results in East Africa. It didn't. Sourcing the transgene from a non-locally prevalent SPFMV strain left breeders locked into a strategy that insufficiently appreciated the agro-ecological links between varieties and viruses specific to a particular geographical region.[59]

Many of these same issues that plagued the GM sweet potato program were evident in IRMA as well. The starting point for the IRMA project was the genetic constructs donated by the University of Ottawa. But after years of testing different Cry constructs and then stacking these Cry events together, breeders were still unable to achieve sufficient resistance to the most pernicious of the East African maize pests. The logic underpinning this project was that the technology had worked well in America, and thus should work well in East Africa, too. But the agro-ecological differences between the United States, where this technology was originally developed, and Kenya, where it was destined to be transplanted, were too great to overcome. Once the possibility of using open-source Bt events was abandoned, IRMA breeders immediately switched gears and approached Monsanto about using their proprietary MON 810 event, though these negotiations eventually faltered. IRMA was committed to using an existing gene construct to mitigate insect damage. Indeed, IRMA's greatest successes were only achieved after it abandoned its commitment to GM, and recommitted itself to providing a series of insect-resistant improved varieties that were well-suited for the full range of agro-ecological growing conditions. This more inclusive breeding approach that moved beyond the narrow focus on transplanting existing Cry proteins was celebrated by IRMA's lead breeder at the project's concluding conference: "We need a more targeted breeding program that incorporates drought, nitrogen use efficiency and maize lethal necrotic disease tolerance."[60] In short, he advocated for an approach that begins with the full spectrum of limitations faced by smallholder farmers, as opposed to an experimental program that hinges on an existing technology that targets a single trait.

These two precedents also focus attention on the critical issue of selecting the host variety to undergo genetic modification. What was particularly innovative about the IRMA project was its initial decision to focus on generating a GM version of an OPV, which is preferred by the vast majority of small-scale growers across the continent. Yet this part of the program was shelved at the same time as IRMA admitted its inability to successfully integrate publicly sourced Cry protein, though the program continued its commitment to breeding for insect-resistant OPVs via conventional means. IRMA remains the only experimental program on the continent that has ever attempted this lofty goal of creating a GM OPV. Most breeders remain averse to this possibility, arguing that it is impossible to maintain the integrity of the GM trait over time because it clashes with farmer preferences for replanting, mixing, and sharing seed. As Willy De Greef, former head of regulatory affairs at Syngenta put it, "the central problem with OPVs as a vehicle for single gene resistance is that it is impossible to keep the varieties pure, especially as it is the stated intention that farmers will recycle the seed."[61] This crucial focus on the choice of host variety was significant within the GM sweet potato program as well, with critics noting the insufficient commitment to developing clones suitable for each of Kenya's distinct agro-ecological zones as evidence that the GM sweet potato program failed to sufficiently recognize the diversity of varieties that farmers use to meet their agronomic and nutritional needs.[62]

One final lesson learned relates to the P3s that managed these breeding programs. Both projects created model P3s that were instrumental in building momentum for the much larger wave of GM staple crops that was to follow. Both partnerships successfully leveraged private sector investment alongside public expertise. But both neglected a paramount dynamic that contributed to their respective downfalls: the delivery and adoption of these technologies. Both projects presumed that an efficient and effective seed system was in place that would enable farmers to access these technologies once they were ready for commercialization, but neither invested in these key arenas of distribution, extension, or education. As the World Bank has emphasized about P3s more generally, "Stronger collaboration between research and advisory services, in particular at the local level, would be needed to improve adaptation of existing high-yield varieties to local conditions and to make existing varieties more profitable for farmers." An earlier World Bank report was even more blunt in its assessment of this missing interface between researchers and farmers: "farmers' needs do not sufficiently drive the orientation of research and extension efforts."[63]

Both the GM sweet potato program and IRMA were examples of P3s that prioritized the potential application of genetic constructs over the complexity of variables that determine farmer well-being. Both put the solution before the problem, anchored by the potential benefits gained by transplanting Monsanto's coat protein, conferring virus resistance (in the case of sweet potato), and the Cry proteins, conferring insect resistance (in the case of IRMA). Both fell into the trap for win-win partnerships outlined by Scoones and Thompson in which "the politics get obscured and, consequently, the underlying tussles over framings, interests, and distributional outcomes become overwhelmed by the focus on technical options and simplistic economic arguments."[64]

A more demand-driven model of experimentation might have started with the full interplay of factors causing low yields in each crop: for sweet potato this would include the tripartite complex of diseases, pest damage, marketing issues, and deteriorating planting materials due to farmers' heavy reliance on recycling and replanting tubers, while for maize this would include limitations around replanting, bolstering existing cultural methods used to control insects, and post-harvest storage, which accounts for both massive losses and devastating health consequences throughout the country via the proliferation of afla-toxins. As Odame, Kameri-Mbote, and Wafula argued in the case of GM sweet potato, a broader examination of farmer priorities could have resulted in a more user-relevant technology that addresses farmers' wide agronomic needs instead of revolving around the transplantation of an existing technology to address a single production constraint in isolation.[65]

CONCLUSION: THE PROBLEM WITH PARTNERSHIPS

A complete analysis of the legacy of these two crucial precedents would be incomplete without an account of what each P3 *did* accomplish. As James Ferguson observed over twenty years ago in his seminal book *The Anti-Politics Machine*, agricultural development projects are often best understood not by what they intend to accomplish, but rather by what they actually accomplish. Viewed within this lens, both the GM sweet potato program and IRMA made important contributions to the long-term prospects of GM crops in Kenya. Both projects made sizeable investments in infrastructure: the GM sweet potato program facilitated the construction of a containment laboratory and the "refurbishing" of other biotech capacity, while the IRMA project was the catalyst for a brand new laboratory, greenhouse, and field trial facilities.[66] Both projects also invested heavily in capacity building and together offered specialized training in transformation and gene construction, training in laboratory protocols, and targeted mentorship programs on skills such as scientific publishing.[67]

Most crucially, these two projects catalyzed the formalization of the country's biosafety regulatory apparatus, which is a prerequisite for the commercialization of GM crops, since licence holders are unwilling to operate without regulations and institutions that conform with international standards. The GM sweet potato program was particularly important here, as it precipitated the country's major regulatory building blocks, including the drafting and adoption of Kenya's biosafety guidelines and the establishment of the NBC; indeed, the National Council for Science and Technology only completed their biosafety guidelines a few months before the NBC approved the importation of the GM sweet potato, which proved to be the body's first regulatory decision. One researcher who followed the project extremely closely observed that the original experiments were deliberately located in the US to put pressure on Kenyan regulators; the program "put conditions in place so that the fruits of this experimental

program can be utilized once the work abroad is completed."[68] This
has led to some critics dubbing it a "Trojan potato," a project that
was never intended to succeed but whose intention was to lay the
groundwork for future biotechnology interventions in the country.[69]
The IRMA partnership that brought together KARI, CIMMYT, and
Monsanto ultimately failed, but it paved the way for future P3s that
sought to create insect-resistant varieties using Monsanto's MON 810
event, a collaboration that ultimately led to the much heralded Water
Efficient Maize for Africa project (discussed in the following chapter).
In this sense, then, the GM sweet potato program and IRMA served
as technology enablers that formalized the regulatory approach and
established collaborations that created a "favourable environment"
for the future entry of new GM technologies.[70]

Viewed within this lens, both the GM sweet potato program and
IRMA exemplify the logic that guided much of agricultural development throughout the twentieth century: "Rural people are reliant
wholly on agriculture, yields are dropping purely because of an environmental problem, a technical solution is sought, and development
will inevitably follow."[71] These two programs were carefully selected
and planned: focusing on staple crops created a sense of ownership
and pride around the project, a sense that "this was our program, this
was our crop."[72] Kenya was identified as an ideal destination for this
investment in biotechnological research precisely because its national
research organization was enthusiastic towards biotechnology but
the country lacked the legislative framework to achieve this progress.[73]
Florence Wambugu, the GM sweet potato's greatest champion, has
argued passionately that this program succeeded in galvanizing political support for GM: "Kenyan and other African scientists have joined
the current debate of biotechnology, actively defending the technology
against the anti-biotech lobby groups that are currently fighting to
destroy the technology ... SSA requires a critical mass of trained
scientists to operate a laboratory-based initiative to develop, transfer
or modify new technologies from elsewhere to make them suitable
for African needs and environment."[74] If an epitaph were to be written on the tombstone of these first two examples of GMO 2.0, it might
be this: the GM sweet potato program and IRMA were scientific
failures, but they were a political success.

Water Efficient Maize for Africa in Kenya

Maize is unique within the category of orphan crops, in that it is consumed widely outside of s s a, it is grown by largeholder as well as smallholder farmers, and it has received ample investment from private agribusiness in order to improve both productivity and profits. Still, maize qualifies as an orphan crop because of its central importance within the continent's food profile: maize covers more than thirty-five million hectares across s s a, with aggregate production exceeding seventy million tons in 2014.[1] Maize accounts for more than a third of the continent's cereal production and just over 8 per cent of the value of total primary crop production.[2] This heavy dependence is most pronounced in east and southern Africa, where maize accounts for 45 per cent of all calories derived from staple cereals, providing more than 50 per cent of total calories in some countries, including Lesotho, Malawi, and Zambia.[3] Across the continent, maize accounts for between 22 and 25 per cent of carbohydrate staple consumption, making it the single largest source of calories across Africa.[4]

But as Bill Gates emphasized in his foundation's 2015 annual letter, maize is the poster crop for the yield gap in Africa. Yields for American maize farmers are more than five times greater than those achieved by their African counterparts.[5] Yields across the continent averaged 1,442 kilograms per hectare in 2014, a minimal increase from the 1960 average of 802 kilograms per hectare.[6] While overall growth in annual production rates from 1960 to 2008 averaged 3 per cent, 2 per cent of this increase came from expanding maize onto new lands, while only 1 per cent came from increases in yield. These yield increases over time rate even more poorly when compared with other growing regions around the world: annual growth yield during this same period averaged 2.4 per cent in Brazil, 1.8 per cent in India, and 2.8 per cent in the Philippines, nearly double what SSA was able to achieve over the same time span.[7]

In 2015 the BMGF announced its big bet to increase the continent's maize productivity by more than 30 per cent by 2030, via investments in fertilizer use, crop rotation, and education around proper planting techniques. A major element of this push is a focus on improved seed, especially targeting the twin challenges of drought tolerance and insect resistance. This chapter chronicles the BMGF's showpiece investment in increasing maize productivity: the Water Efficient Maize for Africa (WEMA) project. It evaluates the evolution and potential trajectory of this project in Kenya, where the population derives 42 per cent of its daily caloric needs from maize.[8] This chapter offers a more cautious view of WEMA's potential in Kenya. It emphasizes the incongruencies between the project plans and the daily realities encountered by its intended beneficiaries, arguing that crucial issues of scale, pest management, credit availability, and gender threaten to undermine this project that holds so much promise for its supporters.

BREEDING FOR DROUGHT TOLERANCE

The African Agricultural Technology Foundation (AATF) is an international not-for-profit organization based in Nairobi that bills itself as a "one-stop shop that provides expertise and know-how that facilitates the identification, access, development, delivery and utilization of proprietary agricultural technologies."[9] AATF first emerged as a partnership between the Rockefeller Foundation and several biotech developers. The Rockefeller Foundation was frustrated by the limiting intellectual property rights regimes that had thwarted their previous attempts to bring new biotechnologies into the public domain. To overcome this hurdle, the Rockefeller Foundation reached out to industry representatives and proposed a partnership that would facilitate restricted public use of proprietary technology. They were able to recruit several major biotech companies for this venture based on the promise of zero liability, as well as the assurance that intellectual property rights would be respected and protected through licensing agreements that would afford them control over their technologies. This partnership led to the emergence of AATF as a facilitator designed to access emergent proprietary technology to address the constraints facing small-scale farmers in SSA.[10]

The AATF was launched in 2004 by Gordon Conway, the former president of the Rockefeller Foundation. According to Conway, AATF was designed to help "unjam the logjam" of intellectual property rights that prevents poor countries accessing and using these technologies,[11] and act as "a focal point where Africans can access new materials and information on which technologies can be built."[12] The organization's initiatives are anchored through its simultaneous connections with farmers and technology developers. It conducts extensive consultations with both groups to determine what technologies are best suited to specific constraints. Once key partners and technologies have been identified, the AATF opens negotiations with technology developers, with the aim of establishing a licensing agreement. This step involves the development of comprehensive risk assessments and indemnification provisions to ensure that liability and intellectual property rights protections are embedded in these licensing agreements.[13] The AATF also engages in monitoring and compliance activities to ensure these requirements and regulations are abided by.[14]

The AATF breeding program with the most financial and political backing is WEMA, which is committed to using genetic modification to produce drought-tolerant maize. As we saw in chapter 4, there is a long history of maize breeding on the African continent that was initiated under colonial experimental programs. But most of these efforts focused on increasing yield, changing characteristics to make them more amenable to industrial processing and mitigating diseases (such as rust). Because the target recipients of these breeding efforts – white settlers living on large estates – generally had the means and technology to access water in sufficient quantity, the trait of drought tolerance was never prioritized within colonial agricultural programs.

Drought tolerance emerged as more of a priority within post-independence breeding programs. In the 1960s, KARI began developing early-maturing varieties designed to minimize exposure to drought, but corresponding lower yields limited the effectiveness and penetration of these varieties.[15] A major breakthrough came in the 1990s, when breeders at CIMMYT piloted a novel methodological approach called "managed stress conditions," which sought to develop varieties that matched up with representative rather than optimal growing conditions, as had been done in the past.[16] The success of these new varieties enhanced the profile of these breeding programs and generated extensive donor interest, leading eventually to two sister programs: Drought Tolerant Maize for Africa (DTMA) and WEMA.

DTMA began first in 2006, with funding from the BMGF and the Warren Buffett Foundation, and coordinated by the CGIAR system. This project aimed to address maize breeding in drought conditions by creating varieties for mega-environments via conventional techniques, and then transferring these to National Agricultural Research Systems (NARS).[17] Its sister program, WEMA, emerged two years later in 2008. While both programs focus on targeting drought-resistant maize by creating a technology pipeline that seeks to connect end users with the breeding process, they differ in organizational structures and methodological protocols: WEMA is a P3 and is explicitly committed to using genetic modification alongside conventional breeding to create a scale-neutral, pro-poor technology that can be adapted to any region.[18] This project is coordinated by the AATF and funded by the BMGF, the Warren Buffett Foundation, and USAID. Institutional partners include CIMMYT and Monsanto. The AATF has mainly taken on a leadership role, while CIMMYT has provided maize varieties, expertise in breeding, and their connections to research programs and seed companies.[19] Monsanto donated their germplasm royalty-free, as part of their Sustainable Yields Initiative. At the national level the project is partnered with NARS, which plays a central role in adapting and distributing WEMA seeds across five partner countries: Kenya, Mozambique, South Africa, Tanzania, and Uganda.[20] DTMA released 230 varieties of drought-tolerant maize in target countries between 2007 and 2015.[21] In 2016, DTMA was absorbed into the broader Stress Tolerant Maize for Africa (STMA) program, which seeks to integrate drought tolerance alongside other principal obstacles including disease and pest tolerance.

WEMA – which is a serendipitous acronym, as it literally translates as "goodness" in Kiswahili – was inaugurated in 2008 and is currently being implemented in two phases. The first phase ran from 2008 until 2012, supported by investments of $39.1 million from the BMGF and $7.9 million from the Warren Buffett Foundation.[22] This initial phase was focused on setting up preliminary working teams and identifying trial locations, as well as initiating conventional breeding programs committed to developing drought-tolerant hybrids. In 2009, Monsanto transferred its patented cold shock protein B (MON 87460), marketed as DroughtGard. This gene was originally identified in a bacterium exposed to cold stress, improving the plant's ability to conserve soil moisture and thereby reducing yield loss from drought conditions.[23] This enhanced drought tolerance is designed such that there is no impact on yields under conditions of normal or heavy rainfall, but it can help to mitigate yield losses during times of moderate drought. Permit approvals and trials began in 2010 with initial harvests taking place in 2011 in Uganda and Kenya.

Breeding for drought tolerance is notoriously tricky. Partly this is due to the trait itself: droughts vary greatly in terms of intensity, frequency, and timing. Also, droughts are in large part defined by their interactions with other abiotic stressors, especially soil quality and temperature, making it difficult to imagine a single-trait breeding solution that could withstand the full range of water stress situations. As such, the trait of drought tolerance has been classified as one of low heritability, meaning there is little chance that genetic improvement will translate into phenotypic enhancement, given the high variability in environmental factors that ultimately determine the trait's expression. Then there is the genetics of the trait itself, which relies heavily on interactions between different genes, as opposed to more straightforward traits such as insect resistance or herbicide tolerance, which follow a linear one-gene code for one-trait mode of inheritance.[24]

For these reasons drought tolerance presents a particularly complex problem for maize breeders. The successful transplantation of cold shock protein B into maize was a huge step forward, with preliminary results showing reduced yield losses exceeding 20 per cent during periods of acute water shortage.[25] But competing evaluations suggest that these yield benefits were overblown: a more critical reading of DroughtGard's effectiveness suggests that it confers only minimal reduced losses in times of moderate drought, outperforming controls by about 6 per cent, which compares poorly with gains made in breeding for drought resistance via conventional means.[26] Indeed, the environmental assessment report prepared by the USDA permitting the release of MON 87460 includes the caveat that "the reduced yield-loss phenotype of MON 87360 does not exceed the natural variation observed in regionally adapted varieties of conventional corn."[27]

Because drought tolerance is a complex trait with significant genetic interconnectedness and heterogeneous environment expression, the WEMA program decided early on to incorporate other biotechnological techniques alongside genetic modification. The goal was to use marker-assisted breeding to produce a suite of conventionally bred hybrids that are well adapted to the varying agro-ecological conditions of each participating country, and then transplant the cold shock protein into each of these to offer maximum protection against drought. This commitment to building upon centuries of improved breeding in local cultivars is important, since maize variety selection across SSA varies considerably: one recent survey found that nearly 70 per cent of all cultivars planted occupied less than 1 per cent of the total area.[28] WEMA's first round of conventional hybrids, known as DroughtTego, were released first in Kenya in 2013. Anecdotal evidence of their performance suggests these drought-tolerant hybrids more than doubled the yield outputs of traditional varieties during periods of acute drought.[29]

In 2011, a key strategic decision was made to incorporate insect resistance alongside the targeted trait of drought tolerance.[30] In announcing this decision the head of the Kenya Agricultural and Livestock Research Organization (KALRO) emphasized the damage wrought by the stem borer, which reduced aggregate maize yields by 13 per cent annually, equivalent to the amount of foreign-sourced maize imported into the country each year. She reported that preliminary WEMA trials with the additional Bt trait had shown excellent results against two of the major stem borer pests in Kenya – the spotted stem borer (*Chilo partellus*) and the African stem borer (*Busseola fusca*) – with a recorded yield increase of 3.72 tons per hectare above the best commercial hybrid.[31]

According to one senior scientist who was part of these conversations, stacking insect resistance and drought tolerance together was designed to "increase the utility of both [traits]. And much more than that, there is evidence that the maize that could be weakened by drought suffers more from insect pest damage."[32] Both GM traits were inserted into conventionally bred hybrids also produced through the WEMA program.[33] As with the gene for drought tolerance, the gene for insect resistance was donated royalty-free, meaning farmers will not have pay fees for use.[34]

The second phase of WEMA began in 2013 and ran until 2018, supported by investments of $48.9 million from the BMGF and $7.5 million from USAID. The aim of phase two is to obtain permissions for further trials and commercial release, and begin stacking the two traits into viable varieties.[35] Although they have yet to commercially release transgenic WEMA varieties, they have begun stacking traits to produce insect-resistant, drought-tolerant maize.[36] It is projected that the first stacked varieties from the WEMA project will be available in South Africa in 2020 followed by Kenya and Uganda, and then by Mozambique and Tanzania.

Preliminary assessments suggest that with this maize, WEMA will be able to accomplish the goals of developing a GM variety that is both insect resistant and drought tolerant, and available royalty-free to smallholder farmers across Africa. One recent internal assessment suggests yield increases on the order of 20 per cent.[37] Another claims that potential yield increases could be as high as 35 per cent within the next five years, which would translate into an additional two million tons of maize that could be saved from the ravages of drought.[38] The most recent data from CFTs suggests yield increases around the order of 29 per cent (Bt hybrids gave grain yields of 9.8 tons per hectare, compared with non-Bt hybrids at 6.9 tons per hectare). Despite only using a sample size of eighteen plants (seven Bt maize plants and eleven controls), these authors conclude that "the study demonstrated that MON810 was effective in controlling *B. fusca* and *C. partellus*. Bt-maize, therefore, has great potential to reduce the risk of maize grain losses in Africa due to stem borers, and will enable the smallholder farmers to produce high-quality grain with increased yield, reduced insecticide inputs, and improved food security ... Safeguarding maize yield through stem borer control with Bt-event will have a huge yield benefit to both smallholder and large-scale commercial maize farmers in Kenya and elsewhere in SSA."[39]

Socio-economic modelling has sought to translate these anticipated yield benefits into financial returns. Economic models suggest that inserting the Bt gene alone should translate into sizeable gains for farmers: one suggests the savings could be on the order of US$25 to $60 million annually,[40] while another suggests aggregate gains could be on the order of over US$200 million over five years with a cost of only $5.7 million.[41] Ex ante assessments of the potential benefits offered by GM drought-tolerant maize estimate returns of US$63 million annually, though with a caution that one-third of these revenues could be captured by private seed companies.[42] These researchers further undertook a retroactive assessment of what Kenyan farmers would have gained had they adopted drought-tolerant varieties during the 2006 to 2016 period, which they pegged to be in the range of US$46 to $78 million.[43] The most recent estimate of the potential

benefits that could accrue from the stacked traits put these at
US$117 million over ten years, with more than 80 per cent of these
benefits accruing to farmers.[44]

For proponents, WEMA exemplifies the potential that P3s create
by combining private sector proprietary technology, public sector
germplasm, and expertise from both: "[The Water Efficient Maize for
Africa project] is bringing some of the best of Monsanto's transgenic
technologies, some of their other forms of biotechnology in terms of
molecular markers, some of their best international germplasm, and
they are cooperating with five countries in Africa – Kenya, Uganda,
Tanzania, Mozambique, and South Africa – to develop a whole suite
of drought-tolerant hybrid maize varieties that are being made available the smallholder farmers."[45] The executive director of AATF boasts
that WEMA serves as a template for how private and public sector
actors can come together to benefit poor, smallholder growers: "WEMA
continues to be a success because of the combined and dedicated
efforts within the partnership: the national agricultural research systems, CIMMYT, Monsanto, and AATF. All these partners have continued to work together, celebrating project gains and resolving any
challenges together for the good of the larger goal, and a promise to
smallholder farmers, a promise of food security and better livelihoods."[46] In spite of all this enthusiasm, there remains a dearth of
research that evaluates whether the technology and prescriptions
associated with the WEMA project will sync with the farming realities
of smallholder growers. The following section aims to put the needs
of smallholder maize farmers front and centre by evaluating the full
spectrum of factors – ecological, economic, and social – that will
determine whether WEMA is able to achieve the lofty goals articulated
by its supporters.

WEMA AND SMALLHOLDER FARMING SYSTEMS

Opponents have dismissed WEMA's potential based on its top-down
approach, which in their view "maintains an established international
division of labour agricultural research, in which international centres
(such as CIMMYT) develop materials for transnational 'mega environments,' which NARS (and other national partners) then adapt for a
broader range of nationally-defined agro-ecological zones."[47] In a
similar vein, Whitfield et al. argue that the WEMA program is beholden
to "reductionist inclinations" that "contain inherent assumptions

about the homogeneous nature of its target farming systems,"[48] which have the effect of simplifying complex farm systems down to core generalizable elements. This critique of WEMA as a continuation of the decades-long tradition of top-down agricultural development seems a bit exaggerated: the program does operate semi-autonomous breeding programs designed to offer site-specific varieties that correspond with local agro-ecological conditions. As of 2017 they had released upwards of ninety different conventionally bred varieties designed to respond to the particular conditions of farmers in Kenya, Mozambique, South Africa, Tanzania, and Uganda.[49]

In particular, Whitfield and his colleagues challenge the assumption of scale-neutrality by zeroing in on two maize-farming communities within the project's target agro-ecological zones, which underline how the close proximity of neighbouring farms will make it difficult for smallholder growers to follow the mandated management practices that are required to ensure the long-term viability of Bt resistance. The major issue is that of minimum refuge areas, which are required to mitigate the inevitable build-up of resistance among stalk borer insects. The close proximity of neighbouring farms was a major reason that resistance level spiked among Bt growers in South Africa, virtually none of whom successfully executed the recommended spatial isolation distance of 400 metres between GM and non-GM varieties, which Monsanto's user guide recommends as the minimum to slow down the buildup of resistant pests.[50]

The impossibility of implementing these recommended refuges is apparent to anyone familiar with smallholder maize farming systems in Kenya. Farms of less than two hectares invariably abut one other; it is nearly impossible for an outsider to distinguish where one farmer's maize crop ends and the neighbour's begins. This close proximity stems from a long-term process of increasing population density and decreasing land size: according to World Bank statistics, arable land per capita shrunk by 40 per cent between 1980 and 2005, from 0.23 hectares per person to 0.14 hectares per person.[51] But, as we saw with Bt maize in South Africa in chapter 4, strict adherence to refugia is vital to preventing field-evolved resistance, defined as a "genetically based decrease in susceptibility of a population to a toxin due to exposure."[52]

A transect survey of one hundred lowland maize farmers underlines the incongruity of implementing these mandated refuge areas within a landscape of increased densification: according to their findings, mandating spatial segregation at even one hundred metres – only 25 per cent of the recommended prescription – would result in farmers losing half of the economic benefits associated with Bt maize due to the additional costs of co-existence; that is, production losses stemming from the mandated buffer zones.[53] Not surprisingly, they identify farm size as a key determinant here, "with larger farms better able to capture potential benefits of Bt maize than smaller ones," as it will be easier for larger farmers to sacrifice some land for buffer zones than it will be for smaller farmers who depend on each and every part of their plot to provide. These researchers recommend setting isolation distances at less than fifty metres for "meaningful adoption to occur," a distance that is one-eighth the recommendation from the technology provider Monsanto.[54]

WEMA scientists are aware of these spatial challenges, and have begun designing strategies to ensure proper management practices within this reality of close proximity of neighbouring farms. One high-ranking agronomist suggests that WEMA seeds could be distributed in packets that contain conventional seed at the top, which is known as seed mixing, or the "refuge in a bag" system. This would ensure that any farmer who purchases Bt seed would automatically implement a refuge (though not one that is spatially segregated).[55] Other experts argue that most places in Kenya have sufficient non-improved varieties to satisfy the requirement for a 20 per cent non-Bt minimum, creating unintended refuge areas, though the question here is whether these are sufficiently distributed to serve as refugia, as well as whether these thresholds will be maintained if adoption rates of WEMA maize are high.[56] Still other experts suggest that other grasses could serve as refugia: the hosts that come up most often in the Kenyan context are Napier grass, a quick-growing feed for livestock that is grown heavily throughout the country, and wild sorghum, which is also grown as fodder throughout the country.[57] Each of these alternative refuge strategies hinges upon the same principle: resistance will be controlled as long as smallholder African farmers grow a certain proportion of their area with non-Bt crops, which, when accompanied by "an active decision by each farmer and strict monitoring," should be sufficient to mitigate the onset of resistance.[58]

None of the three strategies outlined above will be sufficient to mitigate rising levels of resistance. The most crucial requirements for any refuge are that it can produce sufficient numbers of high-quality moths, that it exceed the minimum threshold required in terms of coverage relative to the area under the Bt crop, and that moths emerge at the same time on both the Bt and the non-Bt crop.[59] Existing research suggests that each of the alternatives listed above will have difficulty attaining this threshold. Studies on *Busseola fusca* reveal that these larger moths migrate over longer distances and for a longer part of their lifecycle than Bt-targeted pests elsewhere. These high levels of larval dispersal and exchange from non-Bt plants to Bt plants expand the risk of exposing larvae to sublethal dosages of Bt proteins, which in turn accelerates the rate of resistance evolution.[60] The use of unstructured refugia and wild host plants as reservoirs are not accepted practice in insect resistance management (IRM), but are often suggested as suitable alternatives in the African context. There is an overwhelming amount of evidence negating the value of utilizing wild grasses as potential refuges. The major obstacle with host grasses such as Napier grass and wild sorghum serving as refuges concerns the low survival rate of larvae, which have been recorded as under 10 per cent, as compared to between 20 to 30 per cent on cultivated crops such as maize.[61] The moths that do emerge from wild grasses tend to have lower weights, lower fecundity, and growth and development cycles than those produced on maize. What is more, while wild grasses are grown extensively throughout the continent their overall cover abundance is insufficient to sustain sufficient numbers of viable moths. Taken together, these factors lead to poor survivor rates, a low carrying capacity, and, ultimately, a low utility refuge.

Given these limitations, the unintended refugia system is most likely to be favoured once WEMA maize is released: the logic being that "there would be enough OPVs or non-GM varieties growing in that particular area to supply the number of moths the refuge is supposed to supply."[62] The challenge here is that this puts the WEMA technology into an uncomfortable catch-22: the more farmers who choose to adopt WEMA, the smaller these unintended refuges become. High adoption rates would undermine the viability of these default refugia. Adoption rates would need to be monitored carefully to ensure they did not surpass a certain threshold, after which this resistance management strategy would no longer be viable.

The question of resistance is even more fraught given recent findings that question whether field-evolved resistance within *Busseola fusca* is a recessive trait. The high-dose/refuge strategy is predicated on the condition that a genetically based decrease in susceptibility to Bt toxins is inherited recessively, as explained by Campagne et al.:

If susceptible alleles (S) in the pest are dominant and rare resistant mutants (R) are completely recessive, then rare resistant individuals (RR) emerging from Bt plants will mate preferentially with susceptible individuals (SS) emerging from refuge plants. Crosses between (RR) and (SS) parents yield (RS) progeny, so if the dose of Bt toxin expressed is high enough to kill 100 per cent of heterozygous (RS) larvae, the HDR strategy should strongly delay evolution of pest resistance to Bt toxins.[63]

But recent experiments suggest partial dominance of resistant alleles, which translates into a higher number of resistant individuals within the population and in turn accelerates the pace of resistance evolution.[64] The end result is a high-dose failure, which can be compensated for by expanding the minimum required size of the refuge. The impact of inheritance on minimum refuge size are significant: one highly cited study in *Nature Biotechnology* estimates that if resistance is completely recessive, then a refuge of under 5 per cent of total land planted is sufficient to delay the onset of resistance for twenty years, but if resistance is partially dominant, the refuge planted would need to be over 50 per cent of the size of the total area planted.[65] Another recent recommendation suggests that refuges for *Busseola fusca* should be above 30 per cent of the total land planted to Bt maize.[66]

Other variables that impact the pace of resistance evolution are tied to the particularities of maize farming systems in Kenya. One major difference throughout much of East Africa is that farmers plant maize over two growing seasons in a single year, as opposed to southern Africa where only one planting is the norm. Kenyan farmers plant during both the long rainfall (March to May) and the short rainfall (October to December). As a result, the insects go through more generations per year, which will serve to accelerate the rate of resistance evolution relative to a single-season growing system as exists in South Africa, a dynamic that is acknowledged by most breeding experts but remains absent within the existing fanfare championing WEMA's potential to help Kenyan farmers.[67]

INPUTS, CREDIT, LABOUR, AND GENDER

These questions about WEMA's compatibility with Kenyan smallholder farming systems need to extend beyond simply farm size and take into account the growing requirements that will accompany WEMA's dissemination, with a particular focus on whether smallholder farmers are able to achieve these, enabling them to take advantage of any yield benefits derived from the traits of drought tolerance and insect resistance. Like any new GM variety, purchasing WEMA seeds on their own will be insufficient for success: farmers will be required to purchase a "biotechnology bundle" of inputs that are critical to the technology's success. An official from AATF explains that "we will promote a package. You will not have benefit of that product without having that package. The package includes fertilizer, proper weeding, proper planting, timely planting, how many seeds you put per hole, what is the plant population you expect ... the seed performs as best as possible with all these in place, then you get your optimal production."[68] This section examines each of the required inputs associated with WEMA – fertilizer, labour, credit – and assesses how available and accessible these are to smallholder maize farmers in Kenya.

Let us deal with fertilizer first. According to the head CIMMYT maize breeder, the standard recommendation for WEMA maize grown in Kenya is expected to be set at forty kilograms of nitrogen per hectare for lowland areas and eighty kilograms of nitrogen per hectare for highland areas.[69] It seems highly unlikely that such requirements will be within reach for most smallholder farmers. Fertilizer use among smallholder maize growers has been trending upwards for some time: one study found that the proportion of maize farmers using fertilizer increased from 55 per cent to 70 per cent in 2007. But this still amounts to an average application of 18.1 kilograms per hectare, which represents less than 50 per cent of the WEMA recommendation for lowland areas and less than 25 per cent of the recommendation for highland areas.[70] Other studies have similarly identified the high labour costs and lack of availability as major impediments to the adoption of both organic and inorganic fertilizer.[71]

WEMA maize will also require additional labour in the form of weeding to ensure maximum productivity. Recommended frequencies will be four weedings in coastal areas, three in the lowlands, and between two and four in the highlands. Again, current estimates suggest that these aspirational minimums will be difficult for farmers to attain. Nearly all smallholders weed by hand. The existing research is that most farmers weed once or at most twice a season; the first a few weeks after germination, the second once the maize has grown beyond sixty centimetres and is ready for top-dressing.[72] Another recent estimate suggests that most farmers aim to prevent weeds by undertaking tillage prior to planting, with only "occasional weeding after crop emergence."[73] This research concludes that labour remains the primary barrier, as weeding one hectare of maize by hand can take more than fifty days of work.

While WEMA seeds will be sold to farmers royalty-free, officials admit that the additional inputs in the form of fertilizer and extra labour for weeding will represent an additional cost that many farmers will struggle to afford.[74] While there is not yet any formal commitment in terms of the mechanisms or amounts of credit that will be rolled out alongside WEMA seeds, officials remain confident that credit will be part of the introductory package: "If the government wants to increase the maize production, yes they will bring the credit."[75] The pivotal link between credit availability and the adoption of improved seed has been documented extensively in the literature: assessments of the original Green Revolution in South Asia show how those farmers who were unable to access credit largely missed out on the benefits associated with high-yielding varieties, while more recent research on early adoption trends in Africa reveal that credit availability was crucial to supporting uptake of Bt cotton, and that adoption rates dipped significantly once this credit had disappeared.[76]

Credit is the grease that makes the technology work – otherwise farmers will be unable to realize potential gains associated with the WEMA technology. WEMA officials acknowledge that smallholder farmers who are unable to access credit will largely be left out of any benefits associated with WEMA:

> You see, the problem with small-scale farmers is that they're not likely to manage it the best way. They are not going to invest as much fertilizer … For the drought one they may not invest in as much fertilizer, as much as weeding. So that can eat into some of the benefits. But, all things being equal … we develop it in such a way that if they manage the new one the same way that they manage the old, they will be still be able to see the benefit … [but] small farmers may not always reap the benefit that the large farmer will get … They may not be able to afford to manage it in the best way that the large farmers do.[77]

Without a major commitment towards widespread credit availability – a commitment that would need to come from donors invested in the WEMA project, since it is unrealistic to expect African governments to fund such large-scale investments – it seems unlikely that farmers will be able to surmount the sizeable gap between their current levels of inputs and those prescribed under the WEMA project.

In May 2017, I visited with a group of smallholder maize farmers whose situation underlines these potential limitations regarding inputs. These maize farmers were part of a farmers' association in Kiambu County, Githunguri Subcounty, located in the south of the Central Province, a few hours north of Nairobi. Population density is high, with an average of over one thousand people per square kilometre, translating into small land sizes that average between one-half and one acre per household. The area receives a bimodal rain pattern that yields two distinct rainy seasons: the long rains run from March to May and the short rains from October to December. Soils are mainly loamy, well drained, shallow, and support good harvests if rains are timely and persistent. Ninety per cent of families depend on agriculture, with maize as the primary carbohydrate staple. Approximately 70 per cent of farmers engage in dairy farming, with 20 per cent growing coffee or tea as commodity crops.

The farmers I met with were members of the Green Vineyards farmers' association, which had been set up with the support of a local agricultural NGO. All of the farmers applied organic manure to their maize crops when possible, but they all struggled with issues of cost and labour. None could afford inorganic fertilizers. Most intercropped their maize with beans or other crops, though a few of the largest landholders chose to grow maize on its own. All of these farmers chose to grow hybrid seed for market, but still produced their own OPVS – known locally as Gikuyu varieties – for their own consumption. And all of these farmers struggled to access credit; indeed, the major impetus for the formation of this cooperative was the difficulty each member encountered in accessing any credit via formal channels. The cooperative is designed as a "bank without walls," where each member contributes a set amount that allows them to borrow periodically from the collective.

One of these members is David, a young, thirty-five-year-old farmer who is the caretaker for a family plot of 1.6 hectares that is shared among his mother and three siblings, and their children. David is far from a typical farmer: he has paired with a local NGO and invested in a water capture system that allows him to irrigate many of his crops. He also participated in a similar program in which he built a shed and the NGO provided two goats. He has sufficient capital to hire two labourers alongside himself and his mother. And yet when I visited his maize plot in May 2017, I saw a plot divided: one-third of the maize was stunted, with nearly 50 per cent less vegetative growth than those plants on the other side of the plot. The difference was that David did not have enough manure available to fertilize those plants. David's plot was also distinct in terms of weeding: half had been weeded, while the other half was rife with weeds. He relayed that it takes him and his mother around three days to weed the entire field. Other neighbours' plots were completely overrun with weeds: barely any soil was even visible. If a young, innovative farmer such as David struggles with shortages in fertilizer and labour, it is easy to imagine how acute this shortage is elsewhere.

These concerns over insufficient levels of inputs are more pronounced for women farmers, who have been shown to employ lower rates of fertilizer use, labour, and credit than their male counterparts.[78] This is in due in large part to women's more precarious position with respect to land title and ownership, as well as increased barriers to accessing information, extension officers, and services. The concern here is that women farmers will be disproportionately excluded from potential benefits offered by WEMA seeds because of their relatively disadvantageous access to the additional inputs required to ensure its benefits. There is very little official material on the gendered implications of this new technology: the BMGF commissioned an independent social audit undertaken at the University of Toronto, which was tasked with investigating the potential ethical, social, and cultural implications of WEMA commercialization. None of the three reports published pays close attention to the potential gendered implications.[79] This lack of attention to the potential gendered dynamics of WEMA maize is alarming. Most of the promotional material assumes that the benefits associated with WEMA maize will accrue to women similarly as they will to men: the head breeder mused that the combination of drought tolerance and insect resistance will allow women "to spend more time improving the life of the family ... you will see cleaner homes and better fed children."[80] Those few researchers who have actually spent time soliciting reactions from female farmers to WEMA seed highlight the difficulty these farmers encounter in accessing inputs, and the low profit margins women farmers are able to reap, suggesting that women farmers might largely be shut out from potential benefits.[81]

One final point about WEMA's ability to sync with smallholder farm systems in Kenya concerns the cultural methods of control currently used by farmers to curtail stem borer attacks. Farmers in Kiambu County place ash, soil, or mud in the central maize whorl as a means of containing stem borer. Maize breeders acknowledge the value of this method, but explain that it is impossible to be scaled up for larger farmers.[82] This practice has been used successfully by maize farmers throughout Africa, including Uganda, Nigeria, and Cameroon.[83] Other recommended but understudied techniques include crop residue management (i.e., burning or burying crop residues after the maize is harvested to ensure it does not persist as a source of infestation for next season), and using deterrents such as a cow urine, red pepper, and tobacco mix, or detergent.[84]

CONCLUSION

This chapter has exposed the gap between soon-to-be-released WEMA technology and smallholder maize farming systems in Kenya, emphasizing what Whitfield et al. refer to as the "distinction between the assumptions that underpin these large-scale system interventions and the farm system-level constraints and dynamics that determine the way that these interventions are experienced."[85] In particular, this chapter unravelled the mechanism through which the knowledge systems governing WEMA are produced and legitimized, shining a light on issues related to land size and density, and challenges in controlling field evolved resistance, alongside deficiencies in inputs, labour, and credit, which together cast doubt on WEMA's abilities to provide tangible benefits to smallholder farmers. Questions of gender and cultural methods of control further complicate the very linear technology pipeline model upon which WEMA is premised. Surveying the full range of economic, ecological, and gendered implications of WEMA's introduction raises significant concern around whether this technology is appropriate for Kenya's maize agricultural system.

Taken together, these findings suggest a profound incongruity between WEMA's assumptions of scale neutrality and rational adoption on the one hand, and the complex dynamics of smallholder maize farming systems on the other. WEMA's original commitment to developing drought-tolerant maize varieties seems sound: it had the potential to improve farmer livelihoods by providing another option for mitigating a difficult abiotic stressor – another tool in the toolkit, as proponents like to say. But stacking the Bt event on top of drought tolerance was a mistake, as it opened a host of new problems related to insect pest control and agronomic management practices. Now, WEMA maize will be subject to the full gamut of complications that plagued Bt maize in South Africa, from concerns over target pests developing resistance, to tensions over hybrid versus OPVS (see chapter 4).

Despite these concerns, the enthusiasm for WEMA maize is very much on the rise, due primarily to two separate ecological events. The first is recent drought in East Africa. Mark Lynas, one of the Gene Revolution's most vocal supporters, expressed outrage at the stark contrast of WEMA maize being burned as part of the normal biosafety regulatory process while nearby farms were being ravaged by drought, which claimed more than half of Tanzania's maize crop at the last harvest. Tanzanian maize growers, he writes, are "surrounded by misery. Drought has stalked the land for over a year. Farmers watch helplessly as their wizened crops of maize – a staple food for the area – shrivel under the relentless sun." Lynas concludes that, due to the current drought, "Tanzania is desperate for drought-tolerant crops." Lynas argues passionately that the WEMA project needs to be accelerated to mitigate the damage of this latest drought.[86]

The second ecological event is the outbreak of fall armyworm (FAW), an invasive caterpillar from the Americas with a short life cycle that can lay up to two thousand eggs on young maize plants. Once hatched, these larvae burrow into the plant and destroy it. The outbreak started in the continent's southern tip and has since spread to over twenty maize-growing countries. According to CIMMYT the insect is responsible for the destruction of over 300,000 hectares of maize in the past two years.[87] Mitigation measures rely on heavy doses of pesticide applications designed to prevent the larvae from reaching the maize cobs. Early in 2017, reports began trickling out of South Africa relaying that Monsanto's YieldGard II GM maize seed, a newer version of Bt maize, was less susceptible to FAW. The president of the Southern African Confederation of Agricultural Unions was the first to go on record claiming that Bt maize did not see any FAW damage, a claim that was eventually confirmed by Monsanto South Africa.[88] These preliminary reports of resistance have spread quickly across the continent. In an emergency meeting in Nairobi in

March 2017, the AATF's WEMA project manager argued passionately that WEMA was the best option for protecting Kenyan farmers against this pernicious pest: "successful tests have proven that the genetically modified plant is resistant from attack as compared to traditional hybrid varieties ... when the fall armyworm attacked it had negligible effect on Bt variety, while destroying the traditional hybrids used for assessment in the research."[89] In Uganda, WEMA scientists have petitioned the government to fast track the long-stalled biosafety bill, claiming that Bt maize is the only protection against the pest.[90] A NARO maize breeder argues that "the ultimate solution to these pests is to breed for resistance. Fortunately, we have a product that has shown to be effective in the control of the pest. It's up to policy makers to pass the bill allowing farmers to access these products."[91] Bt maize has been advocated for on similar grounds in Zambia[92] and Kenya,[93] leading some experts to wonder whether the threat of FAW will be the impetus for WEMA's ultimate commercialization: "I have often wondered whether this is what's going to provide companies with leverage to convince African governments to move now on this topic that has been stalling for so long with the approval."[94]

7

GM Banana in Uganda

Banana is an enormously important crop in Uganda, accounting for over 30 per cent of the country's daily caloric intake.[1] Over 75 per cent of Ugandan farmers grow banana across 600,000 hectares of land. Thirteen million Ugandans rely on banana as their primary staple crop, which accounts for nearly a quarter of the country's agricultural revenue.[2] Uganda is the second-largest producer of banana in the world with annual production exceeding ten million metric tons.[3] In fact, Ugandans consume more bananas per capita than anyone else on the planet, estimated at 0.70 kilograms per person per day.[4]

My research over the past five years has focused on one of the showcase experimental programs dedicated to using genetic modification to improve African staple crops: the cooking banana in Uganda, known locally as matooke. The East African highland banana (*Musa acuminata*) is distinct from the Cavendish or Gros Michel bananas familiar to most North Americans and Europeans. Unlike these dessert bananas, matooke is picked before it is ripe, and sold by the bunch. Individual fingers are peeled, steamed, and mashed, then wrapped in banana leaves and roasted over a fire for between four to six hours. The end result is a bright yellow mash, similar in consistency to mashed potatoes, that is by far the single largest source of calories in Uganda. A common refrain heard through Uganda is that it there is no meal without matooke (*kati eno mmere ki nga tekuli matooke*).

Matooke's unrivalled importance within Ugandan farming systems – alongside considerable growing constraints and breeding challenges, relayed below – made it an ideal candidate for the application of genetic modification as part of this new wave of enhanced carbohy-drate staple crops. This chapter recounts the trajectory of genetic modification's application to matooke, unravelling the complex arrangement of institutions and organizations that have created one of the continent's showcase platforms for the potential that biotech-nology can offer to orphan crops. It then presents results from a five-year research project that used both quantitative and qualitative methods to evaluate farmer perspectives on these soon-to-be-released varieties against the assumptions employed by breeders. These results offer a more complex assessment of the potential for GM matooke to enhance yields and livelihoods in Uganda: this project is one example of farmer priorities genuinely being taken into account. It is also an example of a project that targets a crop and traits that matter to smallholder growers, though potential hurdles – in the form of tensions around trait prioritization, selection of host variet-ies, and cost – could undermine its ability to enhance smallholder growing systems.

MATOOKE BANANA IN UGANDA

Matooke is omnipresent throughout Uganda. It is grown on nearly every farm in the heavily populated eastern, central, and southwestern regions of the country. It is available for purchase everywhere: at markets, by the roadside, at the farm gate. Banana bunches are con-stantly in transit: packed into massive lorries destined for city markets, on the backs of boda-bodas (motorcycles) and bicycles, on the top of matatus (communal taxis). Most aim to purchase large bunches that, depending on the size of the bunch, could feed a family of six for between three to six days. Others are forced to focus exclusively on their next meal and do their purchasing by individual fingers.

Banana is ideally suited to Uganda's climate. It grows well through-out the tropical cycles of wet and dry seasons. It is a perennial crop, meaning that it keeps producing new fruit constantly: banana takes fourteen months to flower, and another four for fruit to fill, by which point there is also a daughter and a granddaughter already starting to grow. It does not require careful storage. It reproduces clonally, meaning that banana plantations can persist over decades: many farmers are still harvesting individual plants that were sowed by their grandparents or great-grandparents. It is a crop that requires relatively little labour: one common anecdote relays the story of a colonial officer who approached a sleeping farmer and was critical of his laziness, to which the farmer replied cheekily that he was taking care of his banana plantation. A Luganda saying sums up this attitude: *Kyosimba onanya kyilake effoke*, "the banana you plant jokingly is the one from which you'll pick a bunch," meaning that matooke will produce fruit regardless of whether a farmer is diligent about taking care of it.

Despite these advantages, Ugandan banana growers face a number of persistent challenges. Pests and diseases are a constant threat in a tropical climate that offers no period of dormancy. Fusarium wilt (FW), known more commonly as Panama disease, is a soil-borne fungus that blocks the flow of water and nutrients, leading to the collapse of leaves and eventually a splitting of the pseudostem. Matooke banana tend to be more resistant to the various incarnations of FW, while sweet bananas and brewing bananas are especially vulnerable. Black sigatoka is caused by an air-borne fungus and is spread by both insects and spores. Like FW, sigatoka attacks plant leaves, depleting chlorophyll levels making the leaf dry and necrotic. This loss of photosynthetic capacity in turn impairs fruit development. The two most common pests are nematodes, soil-borne worms that weaken the root system, damaging the stalk's stability and making it more susceptible to wind damage, and weevils, soil-borne insects that eat into the stalk directly, damaging the stalk's stability and making it more susceptible to being toppled by heavy wind. The damage wrought by these two pests is cumulative and takes years to notice, as opposed to the diseases whose damage is more immediate and observable.[5]

The most devastating stressor affecting matooke in Uganda is banana bacterial wilt (BBW) disease.[6] BBW is a vascular disease that impacts crop yields, incomes, and livelihoods.[7] Known locally as kiwotoka, BBW was first identified in 2001 in the Mukono District of central Uganda. It is caused by the bacterium *Xanthomonas campestris pv. musacearum*, the same bacterium that caused enset wilt in Ethiopia during the 1960s.[8] Symptoms of bacterial wilt include rapid yellowing and wilting of the plant and leaves, as well as premature ripening of the bunch, discolouration of the pulp and internal vascular vessels, and secretions on the leaves. The disease can spread quickly before exhibiting any symptoms. Once visible wilting occurs, the disease has likely progressed through the plant and can no longer be recovered.[9] It is estimated that BBW costs Ugandan farmers over US$500 million annually.[10]

Research scientists at Uganda's National Agricultural Research Organisation (NARO) have been attempting to breed improved varieties that can resist this collection of pests and diseases, but with little success. Banana breeding is notoriously difficult. Banana has very low levels of male and female fertility, meaning it is essentially sterile. It has a long generation time (eighteen months for first fruiting and then six months for subsequent generations), which makes for a slow and time-consuming breeding process. These characteristics make introducing beneficial traits into popular varieties via conventional breeding extremely challenging.[11]

NARO's conventional breeding efforts have centred on identifying wild races that demonstrate considerable resistance to pests and diseases, and crossing these with preferred varieties that are widely cultivated. The most effective method is to take fertile pollen from inedible wild banana and hand-pollinate matooke fingers to yield viable seed. But the success rate is extremely low, averaging only two or three viable progeny per attempt.[12] Indeed, the most successful improved matooke variety is actually a misnomer. Known as Mpolologoma, this high-yielding variety produces a big bunch and is tolerant to the most common pests and diseases, making it a favourite amongst farmers and consumers alike. But most farmers erroneously assume that Mpologoma is an improved variety produced by breeders. In fact, Mpologoma is a superior landrace that breeders identified through a national collection of banana varieties; the only role breeders played was to propagate and popularize it.[13]

These challenges associated with conventional breeding led NARO to approach the UN's Consultative Group on International Agricultural Research (CGIAR) for an initial investment in GM experimental capability around the turn of the millennium. This support for infrastructure and training expanded in the ensuing years, incorporating commitments from USAID via the ABSP II project, which provided bridge funding to the newly trained PhD plant breeders to ensure their return to Uganda. ABSP II further paid for the construction of a new tissue culture laboratory, a biosafety greenhouse, lab equipment, and fencing, with a focus on bringing these new agricultural biotechnologies "from research to the field."[14]

These nascent efforts centred on creating a GM version of matooke that could resist this cocktail of pests and diseases. While USAID provided the vast majority of infrastructure and capacity building, the BMGF-funded African Agricultural Technology Foundation

(AATF) set about negotiating access to genetic constructs from private companies for licence-free use by NARO scientists.[15] NARO breeders are the first scientists in the world to undertake the genetic transformation of bananas, with over eight separate experimental lines currently in operation. The ultimate goal is to combine all of these GM traits into a single high-yielding variety that will be resistant to all of the region's pests and diseases. Scientists call this variety Banana 21: the banana for the twenty-first century.

I began visiting with the research scientists leading this nascent research program dedicated to developing a GM banana in 2009. Over the next three years I met with dozens of research scientists, development donors, and policy officials asking questions about the experimental design of these novel programs and their potential to help alleviate poverty and hunger for small-scale farmers. Our exchanges about breeding strategies and dissemination techniques were illuminating. But the conversations stalled when I asked about how farmers would respond to these improved varieties. Informants were convinced that farmers would embrace the new GM varieties, despite the fact that farmers had been minimally consulted to ensure their preferences were taken into account. Most dismissed the idea of bringing farmers into the debate because if the technology was not yet available to them, "Farmers cannot understand because they have not seen it."[16] They suggested that farmers only needed to be brought into the conversation once the technology was ready for commercialization, "in the final stages of testing the technology, when we have something in hand, that's when we involve the farmers."[17] Other biotech boosters maligned the amount of misinformation being propagated, arguing that farmers do not have sufficient capacity to distinguish accurate reports from misleading ones. One high-ranking official complained that farmer knowledge around agricultural technology was overblown: "some people think farmers know everything."[18]

In 2012, I embarked upon a long-term research project designed to challenge these assumptions. With funding from the Social Sciences and Humanities Research Council of Canada and the John Templeton Foundation, I assembled an extremely talented group of researchers who canvassed banana growers across the country in order to assess how well these improved varieties matched up with the ecological, economic, and cultural realities they encounter on the ground. We created a random sample of 172 banana growers that was representative of the country's diverse grouping of agro-ecological conditions, as well as categories of power such as gender, ethnicity, age, education level, and landholdings.

The starting point for this sample was the 2008/09 Uganda Census of Agriculture, which revealed that Ugandan banana growers are spread unevenly across the country: 15 per cent reside in the eastern region, 35 per cent in the central region, and 50 per cent in the southwest. We created a random, stratified sample to reflect this geographical distribution. Districts were randomly selected based on the most recent census. A random number generator was then used to select sub-counties, parishes, villages, and, finally, individual households.[19] We ended up visiting with thirty-three farmers in the eastern region, sixty-two in the central region, and seventy-one in the western region. The vast majority of these farmers (over 85 per cent) had not completed formal schooling beyond secondary. Participants were mostly smallholders; the average farm size was 2.2 hectares, with 54 per cent of farmers having fewer than 1.2 hectares. Farmers were generally quite poor, with average expenditures per week of US$15.80 spread out across an average of eight dependents per household. An additional one hundred farmers were invited to ten separate gender-specific focus groups designed to offer participants a chance to comment and elaborate on the results achieved via the on-farm ranking exercises.

Our project combined qualitative and quantitative techniques to generate a methodological program that captures the complexities of farmer decision-making. It bridges the gap between the models and surveys favoured by economists and the ethnographies and participant observation favoured by anthropologists to offer an innovative methodological approach for evaluating farmer attitudes and behavioural intentions to adopt new agricultural technologies. The methodology used in this project emerges out of the conceptual framework of participatory plant breeding, which is guided by a desire to integrate farmers directly into the agricultural innovation process.[20] This progression of exercises relies heavily on visual aids, future scenarios, and side-by-side comparisons, which helps to bridge the gap between hypothetical exercises and farm-level realities. After using quantitative exercises to determine farmer attitudes and intentions, focus groups were used to understand why farmers felt the way they did. Ten gender-specific focus groups consisting of between six to ten farmers each were convened in each of the major growing regions to provide insight into the meaning of quantitative results. Each focus group began by undertaking the same series of exercises described above as a collective, prompting members to question, listen, and disagree as they tried to reach a consensus. Our overarching aim was to assess farmer perspectives on these soon-to-be-released GM matooke varieties.

WHOSE TRAITS? WHOSE VARIETIES?

Our first series of questions surrounded farmer attributes: we were interested in knowing which farmer characteristics (gender, age, education, etc.) shaped farmer attitudes and intentions to adopt these soon-to-be commercialized varieties. Our results – recently published in the journal *AgBioForum* – found that five characteristics were linked to positive attitudes towards GM banana: farm size, geographic location, membership in a farming association, experience with improved varieties, and visits from agricultural extension officers. Basically, the farmers who showed the most enthusiasm towards genetic modification were those with larger holdings, the majority of whom were based in the western region, and those who enjoyed privileged access to information, training, and existing improved planting materials. We concluded with a cautionary policy recommendation: the five variables that significantly impact attitudes and intentions to adopt are all associated with affluence and influence. These results thus raise important questions about the potential for GM matooke to help the poorest and most vulnerable in the country – that is, those who are disproportionately located in the eastern and central region, with smaller farms, who tend to be excluded from formalized social networks and lack critical access to information. These results underscore the need for targeted policies geared towards meeting the needs of farmers who lack these resources, and the risk that these most vulnerable farmers could miss out on the potential benefits associated with GM versions of African carbohydrate staple crops.

Our second series of questions sought to examine whether the priorities for the GM varieties currently under experimentation sync with the priorities and perspectives of the farmers who are their intended beneficiaries. Our first questions were about which traits were being prioritized in the experimental programs and whether these aligned with farmer concerns about which pests and diseases were limiting their yields and livelihoods. To answer this, we undertook a participatory ranking exercise that presented farmers with the full gamut of biotic and abiotic growing constraints: every pest and every disease, alongside other preferential characteristics such as larger bunches, faster growing times, colour, taste, and texture. We then asked farmers to rank which traits were most important to them. Eight of the top nine categories selected by farmers related directly either to pest and disease resistance, such as BBW, FW, and weevils, or the increased yields that should result from mitigating these biotic constraints, such as bigger bunches, faster growth, etc. These quantitative results suggest that farmers are largely supportive of orienting the GM experimental program towards mitigating biotic constraints. Qualitative data gleaned from focus groups corroborate this widespread support for using genetic modification to increase yields. Male farmers in the eastern district of Kamuli expressed confidence in genetic modification as a breeding strategy: "let us embrace this technology," said one, while another opined that "modifications are made to achieve a better product. If genetic modification comes, we will have much better food security."[21] A female farmer in this same district concurred: "[genetic modification] is a good innovation, in that the spread of diseases and attacks by pests can be reduced."[22] A farmer in the western district of Kyenjojo emphasized that they supported genetic modification as a means of mitigating damage by pests and disease: "it is good because scientists have proved it to be the best way of preventing BBW and fusarium." His neighbour expressed confidence that genetic modification would be embraced by farmers desperate for anything to help avert losses to pests and diseases: "it will be good because without them we are likely to lose the bananas completely."[23]

But the first GM variety scheduled for release will address neither pest nor disease resistance. The first variety scheduled for release in Uganda is the biofortified variety, which has been engineered to respond to the country's high rates of hidden hunger. Micronutrient deficiencies in terms of vitamin A, iron, and zinc – known collectively as "hidden hunger" because these insufficiencies are less visible and more difficult to identify than caloric insufficiency – carry with them subtle but severe consequences that include anemia, blindness, skin lesions, stunted mental and physical development, increased frequency of illness, and premature mortality. These ailments are especially pronounced amongst children under the age of five, as well as pre- and post-natal women.[24]

Uganda is typical in terms of the prevalence of hidden hunger throughout the country and the increasing enthusiasm for using biofortification as a means of redressing it. According to the most recent national demographic and health survey available, 38 per cent of children aged six to fifty-nine months and 36 per cent of women are deficient in vitamin A.[25] Biofortification of matooke is advanced as the most promising strategy for addressing this nutritional shortcoming because most of the poor in Uganda are subsistence farmers living in rural areas, making biofortification more practical than alternatives such as supplementation of vitamin pills or the fortification of processed foods.[26]

Over the past decade, biofortification has emerged as the preferred mechanism for addressing hidden hunger across Africa. Breeding enhanced micronutrient content into staple crops has been championed both on arguments of cost-effectiveness and on its high adoption rates within rural areas that previous programs focused on supplements were unable to access. Given the heightened challenges of conventional breeding in banana, the biofortified banana project was initiated in 2005 as a collaboration between the Queensland University of Technology (QUT) and NARO, with funding provided by the BMGF's Grand Challenges in Global Health. Led by renowned biotechnologist James Dale, the team was able to extract a gene conferring high rates of provitamin A from the Asupina banana variety and transplant it into a matooke host. The first laboratory trials were initiated in Queensland in 2007, and field trials were planted at Uganda's Kawanda Agricultural Research Institute in 2010. Preliminary results suggest that the biofortified banana has been able to achieve pro-vitamin A levels that are between two and four times higher than conventionally bred equivalents.[27] Follow-up experiments investigating multigenerational persistence, availability within food preparation methods, and its performance within different agro-ecological conditions were also positive. The most recent estimates suggest the biofortified banana should be ready for commercial release by the year 2020.

In a recent article published in the *Journal of Peasant Studies* we explore farmer preferences for traits that address issues of yield over those that address nutrition.[28] Enhanced nutrition ranked ninth out of fourteen amongst traits prioritized by farmers in the participatory ranking exercises, lower than every single pest and disease. We subsequently asked farmers to explain these preferences. Some farmers stressed the importance of the diseases: their prevalence, their destructiveness, and the lack of resources available to mitigate them. Others minimized the importance of nutrition, arguing that a greater nutritional value does not translate into higher prices when sold at the market, and expressing confidence that their current diet was already meeting their nutritional needs. One female farmer asked rhetorically, "[the traditional variety] already has lots of vitamin A and iron – why do we need this new one?" while a male farmer stated confidently, "I can get vitamin A from carrots and iron from beans."[29] Larger commercial growers in Ntungamo echoed the sentiment that their current diet provided sufficient vitamin A: "we prefer G M O banana that is pest and disease resistant to nutritionally superior because the local varieties have some vitamins."[30] Some farmers, though, were more willing to acknowledge the potential gains offered by a biofortifed variety: one male farmer in Kamuli stated, "we know our banana contains a lot of water." Still, the quantitative results raised here are unequivocal: farmers rank nutritional enhancement as significantly less important than characteristics associated with higher yield and pest or disease management.

These results suggest that a matooke banana that was genetically modified to enhance yield or mitigate effects of pests or diseases would be more positively received by farmers. And yet it is the biofortified banana that has emerged as the flagship program for Uganda's biotechnology sector. While G M versions of disease- and pest-resistant bananas are at varying stages of laboratory, greenhouse, or field trials, none has received anywhere near the amount of publicity or funding dedicated to the biofortified variety. I asked dozens of experts about this disconnect and received three primary explanations. First, banana breeders argued that biofortification was a "proven technology" whose proof of concept had already been confirmed in golden rice, the highly touted biofortified rice that had been the centre of experiments in Southeast Asia for over a decade. One senior breeder explained, "we knew it had worked in golden rice and how it worked, and potentially it was very easy to put the technology together and apply it to banana."[31] Second, this emphasis on biofortification reflects donor priorities: the biofortified banana program was originally funded through the B M G F's Grand Challenges in Global Health program, which targeted funds towards health-based interventions as opposed to agricultural-based constraints. Third, experts routinely downplayed farmer prioritizations, arguing that nutrition was a bigger issue than farmers understood, and that prioritizing the biofortified banana was in their best interests.

When pressed about this disconnect, farmers offered compelling evidence for why disease and pest resistance – especially resistance to BBW, the most pernicious disease currently affecting Ugandan banana growers – should be prioritized as a GM trait. BBW is notoriously difficult to control. It is transmitted via multiple channels, including contaminated tools, infected plant materials, insect vectors, and infected soil. This leads to skyrocketing rates of infection: within the first year of outbreak in 2001, incidents of the disease rose to above 70 per cent across Uganda.[32] Within five years more than thirty million matooke bananas were infected, culminating in losses estimated at just under US$35 million in 2005.[33] These losses swelled to over $75 million just one year later.[34] The disease has now spread to every banana-growing county in Uganda.[35]

Controlling the disease has focused on reforming agronomic practices designed to curtail its spread, such as de-thrashing leaves, excising excess plans, stump removal, sanitizing tools, and leaving infected soil fallow for six months to avoid recontamination. During the height of the outbreak, additional labour was required to remove male buds from infected plants; uprooting, burning, and burying the whole mat of affected plants; and disinfecting all farm tools that come into contact with affected plants.[36] Educational campaigns designed to raise awareness about these beneficial practices helped to hinder the disease's spread in the north where the incidence rate peaked at 63.4 per cent in 2005 before falling sharply to 4.9 per cent in 2010. But these measures were slower to catch on in the south, so the government shifted towards more aggressive containment measures, focusing on area-wide sanitation designed to ensure that all farmers were compliant. The government went so far as to introduce legislation in 2010 setting fines and even jail-time for any farmer who failed to implement these strategies. These heavy-handed measures created a pronounced dip in infection rates, but these recovered swiftly over the following years once enforcement began to wane, as one recalcitrant farmer was sufficient to undermine the collective efforts.[37] NARO's own internal figures on BBW incidence reveal that these measures did little to stem the disease's spread, as it has now reached districts in the northwest (such as Masindi) and in the far east (such as Bukedea), even spreading beyond Uganda's border, infecting banana plants in neighbouring Democratic Republic of Congo, Rwanda, western Kenya, and northwestern Tanzania.[38] The government then shifted towards large-scale community mobilization efforts in 2012, which succeeded in reducing prevalence in targeted areas; in the Ankole region of southwestern Uganda, prevalence rates dropped from 45 to 13 per cent in a matter of months.[39] Again, however, these reductions have been difficult to sustain over time: incidence maps comparing infestations in the country's banana-dependent southwest in districts such as Ntungamo, Isingiro, and Mbarara show that the severity of the infestation has almost tripled from 2010 to 2015.[40]

While the unrelenting advance of BBW has proven nearly impossible to contain, biofortification is a problem that is comparatively easily addressed by other means. Numerous studies confirm the effectiveness and affordability of household strategies to increase the bioavailability of micronutrients.[41] Among the most promising strategies is increasing dietary diversity via household gardening, alongside training and nutritional information on the incorporation of vitamin A-rich foods, with evidence that this strategy significantly increased vitamin A content in Bangladesh, the Philippines, and South Africa.[42] These studies have underlined corollary benefits as well, including surplus food contributing to multiple health benefits and extra income, and enhanced participation of women in household decision-making.[43] Dietary diversification also overcomes some of the most trenchant obstacles associated with biofortification, including access in remote areas and the fact that vitamin A requires a varied diet (especially in fats) in order to enhance bioavailability, which a narrow emphasis on biofortification fails to address.[44]

Dietary diversity is a particularly acute issue in Uganda. The FAO country report for Uganda notes that the dietary diversification index for Uganda is 57 per cent, a score that is relatively high compared to other African countries.[45] However, the FAO report explains that this score is deceptive because of the heavy reliance on banana in Ugandan diets. This is counted as a fruit and vegetable in the index, despite the fact that it has a comparatively low content of vitamins, leading the FAO to conclude that the Ugandan diet is insufficiently diversified. This view is shared by banana breeders, who argue that overreliance on matooke is a leading cause of hidden hunger in the country: "In southwestern Uganda, if you look at the data, if you look at the livelihood in Bushenyi, they eat bananas for breakfast – without anything, with a cup of tea, maybe with milk, maybe with milk if they can – and that's it. For lunch, if they don't have a sauce they just put some salt on it. You know, and Bushenyi … [has the] highest micronutrient deficiency [in the country]."[46]

In sum, BBW is a trait that lends itself to genetic modification more than biofortification: it is a disease that is difficult to control via changes in agronomic practice or policy, that cannot be addressed via conventional breeding, and that has exasperated breeders searching for mitigation measures. And yet it is biofortification that has been prioritized over BBW resistance, a health-related concern with a cost-effective remedy that has been proven elsewhere. This erroneous prioritization of traits underlines the rot within these experimental programs: traits are being determined by donors instead of by the farmers these technologies are supposed to help.

GENETIC MODIFICATION AND BANANA GROWERS' PREFERENCES

A final series of research questions sought to assess whether new GM varieties are compatible with current growing regimes and farmer preferences. Any assessment of whether biofortified bananas can help the poor needs to take account of how the poor define their own interests and priorities. This section builds upon the 2016 analysis by Glover et al. of the adoption problem by unravelling the agricultural system designed to deliver biofortified banana to Ugandan farmers, in order to evaluate whether this GM banana syncs with the realities of smallholder farmers.[47]

Our conversations with farmers revealed a number of potential issues. The first relates to the host variety that these GM traits will be inserted into. As we have seen through the cases of Bt cotton in Burkina Faso and Bt maize in South Africa, the choice of which variety will serve as host for the GM trait is a key consideration that shapes adoption outcomes. There are over eighty-five endemic varieties of matooke banana, and they all look the same to me, whether they are sitting by the roadside for purchase or served on a steaming plate of food. But to Ugandans these varieties are distinct for their attributes related both to production (bunch size, stem strength, finger length, etc.) and to consumption (taste, colour, texture, etc.). Some of the most popular throughout the country are preferred for their hardiness, their large bunches, and their favourable taste, colour, and texture.

The Ugandan scientists in charge of the GM breeding program are well aware of these value-based preferences and set out to create GM versions of the most popular varieties. But technical constraints in the experimentation process, particularly in the stage of embryogenic cell development, forced them to exclude the most popular traditional varieties. The two host varieties that will constitute the first round of GM bananas are mid-range in terms of consumer preference – as one banana breeder put it, "it is not the best but it is not the worst."[48] But the farmers we spoke to complained about small fingers, slow growth, and stalks that were susceptible to wind damage, and that it produces an unappealing gritty texture when mashed. These less desirable attributes have the potential to slow adoption rates, based not on the GM traits themselves but rather on unappealing traits present in the host variety.

In order to address this potential vulnerability, the second-generation GM banana will see GM traits inserted into an improved variety that has been bred through conventional means. The improved variety that is the ultimate destination for the traits of biofortification and BBW resistance is known by many names: banana breeders call it M9, or Kabana 6H, while farmers have given it the name Kiwangaazi, which translates as "the one that lasts for a long time." M9 is a conventional hybrid, the product of a traditional variety, Nakawere, crossed with the Calcutta 4 banana from India, with resistance to black sigatoka, a common fungal disease that attacks banana leaves, leading to fruit that is stunted and spoiled. The resultant progeny was resistant to sigatoka and tolerant to weevils and nematodes, but also had a taste and texture that was largely unappealing, so it was crossed again with a male parent imported from Honduras.[49] This final version of M9 – the product of more than twenty years of conventional breeding – demonstrates favourable consumption characteristics and persistent resistance to most pests and diseases. It is also a high yielder, producing big bunches that can yield anywhere between twenty-five and seventy kilograms.[50] Banana breeders are effusive in their assessments of M9, which they laud as "outstanding."[51] They remain confident that the long-term plan of introgressing the biofortification and/or BBW trait into this improved variety offers the best strategy for addressing the target trait within a variety that has broad appeal due to favourable agronomic and consumer characteristics.

But our conversations with farmers cast some doubt on these effusive assessments of M9's potential among farmers. There are a couple of points to make here. First, farmers note that M9 and other improved varieties – known as Kawanda varieties due to their being bred at NARO's Kawanda Agricultural Research Centre – are less hardy and more labour intensive than traditional varieties. M9 requires more fertilizer, wider spacing, and onerous management practices including regular de-suckering, de-leafing, and removal of male buds, and farmers often lack the funds and labour necessary to fulfill these onerous prescriptions. A second major complaint about Kawanda varieties relates to value-based preferences, particularly taste and colour. M9 has a reputation for being too hard and too brittle, which makes it difficult to mash, and for having a taste that is unpalatable.

But M9's biggest barrier is its colour. The colour of the matooke mash is extremely important to consumers. Farmers emphasized the importance of matooke retaining a deep yellow hue when it is served: "when it is white it is not attractive, but when it is yellow everyone would like to eat."[52] A farmer from Wakiso District elaborated: "when you sell a bunch, and it turns out to be a white variety for cooking, the buyer will not come back to you. That makes you lose the market because the buyer thinks you sold food harvested too early."[53] This emphasis on colour was more pronounced among female respondents. One female farmer in Ntungamo explained that "when the colour is not good, people don't get appetite for food," while a female farmer in Kamuli said, "if the colour of the food is not good, then the husband [will ask] why the colour is like this."[54] Other women relayed that they would get beaten by their husbands for serving matooke that was too white. The issue of colour has been a challenge since the introduction of M9, with breeders acknowledging that the banana's hue has been more golden than yellow, which has been a major turnoff for consumers: "the problem has been actually we have had the high-yielding banana varieties but farmers here have their own preferences for the cooking banana. A banana that is not soft, a banana whose pulp is not yellow, won't be taken up by consumers."[55] One male farmer in Kamuli summed up these concerns succinctly: "if it is white then it is not even matooke."[56]

Farmer resistance to M9's taste and colour have proven so pervasive that it is in the process of being replaced by a new generation of high-performance hybrids improved via conventional means. Known by the acronym Nabios (national biodiversity hybrids), this new generation of hybrids is built upon consumer acceptability: Nabios have been bred to retain the high yields and tolerance to multiple biotic stressors, but have been subjected to a much more rigorous series of sensory evaluation trials in which focus groups of farmers were convened to rank its acceptability based on texture, colour, taste, and smell. Nabios scored above five out of six on each variable, leading to an overall consumer acceptability score exceeding 80 per cent. Breeders are optimistic that this latest generation of hybrids will be able to avoid some of the pitfalls that have plagued M9. As the head of the Nabios breeding program explains, "the ultimate goal is yield. But yield without the farmer preference is nothing. You'll have a very big bunch of banana that will be rejected on the grounds of organoleptic traits."[57]

Another concern around M9 centres on the question of trust. Many of the farmers I have spoken with over the years have expressed cynicism towards Kawanda varieties based on previous negative experiences. Much of this was rooted in the precedent set by the failed Fhia variety. Fhia is a high-yielding banana from Honduras that was imported to Uganda in 1991. Seven different varieties were included in experimentation trials at Kawanda. While the Fhia bananas were actually bred as dessert bananas, a few varieties were specifically marketed as matooke: while consumer acceptance trials revealed persistent negative reviews based on taste and texture, NARO scientists remained optimistic that select Fhia varieties "had ratings that could make them acceptable for cooking in parts of the country where the cooking banana landraces are disappearing."[58] Fhia varieties were released commercially to farmers a few years later, outyielding every existing banana variety. Farmers were impressed with the large bunches and long fingers, and initial adoption was widespread: farmers were so enamoured with the huge bunches of Fhia 25 that they dubbed it "Kilometre 25," because its bunches were exaggeratedly said to be as long as a kilometre.

But production dropped off precipitously only a few years later.[59] Farmers' objections centred on value-based preferences: they observed that it was too hard to mash, that it did not stick together after mashing, that the food darkened when exposed to air, that it became firm upon losing its heat, and that it produced a white, pale colour instead of the coveted deep yellow. Women farmers were the most adamant in dismissing Fhia: farmers in Wakiso described it as "tasteless," while farmers in Kamuli criticized it for not having the right colour. Today, Fhia is grown as a banana-of-last-resort in the eastern, central, and western regions, where it is consumed only in times of desperate need, though it is cultivated more widely in nontraditional banana-growing regions such as northern Uganda, Tanzania, and South Sudan.[60] In the main banana-growing region, a bunch of Fhia can cost as little as 14,000 Ugandan shillings, which is more than 50 per cent cheaper than traditional varieties that yield much smaller bunches, while in the eastern region a Fhia bunch goes for less than a third of a favoured traditional variety.[61] The legacy of the Fhia debacle is a persistent distrust toward improved Kawanda varieties, which could prove a significant barrier, as these will serve as the primary host varieties for all GM traits in the future.

Two other potential barriers emerged from our conversations with farmers. The first relates to cost. While farmers will not have to pay for the licence associated with the genetic transformation of the bio-fortified gene, they will pay for the tissue culture plantlet that serves as its vessel. Many of the farmers willing to purchase plantlets via the formal seed system identified high cost as their primary barrier to entry. A sucker from a neighbour costs as little as 500 Ugandan shillings, while plantlets purchased at nurseries typically cost between 2,000 and 3,000 shillings. These cost differentials serve as barriers to many farmers. One female farmer in Buikwe District told me that "we would wish to buy from nurseries but the cost is high," while another woman in the same focus group confirmed, "the cost of buying planting materials is too high for me."[62] A female farmer from Wakiso District was strategic in her decision-making, arguing that the cost for a single tissue culture plantlet is so high that it makes more sense for her to buy three or four suckers from a neighbour, which will enable her to earn more.[63] Banana breeders acknowledge the high cost of tissue culture as a potential barrier to adoption: "certainly [the high cost] will slow it down a little bit."[64] This is corroborated by other studies, which have shown that larger farmers tend to adopt tissue culture bananas more frequently than poorer farmers with smaller plots.[65]

A second related hurdle concerns the ability for poor farmers to access these improved planting materials. Unlike GM versions of open-pollinated crops with which the loss of hybrid vigour forces farmers to obtain new planting materials on a regular basis, the GM banana will precipitate a blending of the formal and informal seed systems; after obtaining planting materials via private nurseries, farmers will be free to replant, exchange, and sell subsequent suckers amongst themselves. But resistance is inevitable, and at some point – scientists cannot estimate precisely when – GM varieties will succumb to the pests and diseases they are engineered to resist and farmers will be forced to repurchase next-generation plantlets. So how often will farmers need to go back to nurseries or vendors and purchase new tissue culture plantlets to replace those whose effectiveness has expired? This will depend on how quickly the improved GM trait loses its effectiveness.[66] But what we can say for sure is that GM varieties will increase farmer reliance on the formal seed sector.

We undertook another round of participatory exercises asking farmers to choose between these two options: a traditional variety on the one hand, with predictable but lower yields, and a hypothetical GM variety that produced higher yields but required the farmer to go back to the formal seed sector to purchase a replacement variety after three years.[67] Scores across our sample of 163 farmers show that farmers overwhelmingly prefer a variety that offers lower yields but does not require regular replacement.[68] When asked to explain these results in focus groups, farmers focused on the lack of land availability for expansion and the associated value of traditional varieties that could last upwards of twenty years in a plantation. One female farmer in Wakiso District explained her own situation in the following terms: "we have a small piece of land, so if I continue to look after [the traditional variety] well, I will continuously sell." A neighbour concurred: "we are living on a plot of fifty by one hundred. I would prefer [the traditional variety] and continuously care for it."[69] A male farmer in Buikwe lamented the additional labour involved with replacing varieties even every three years: "Replanting is a laborious process: you make the hole, then you fill it with manure. Then you leave it for some time to cool. Then you plant, then you mulch the plantlet. If it is the dry season, you then put in the drip ..."[70] A female farmer in the same district shared this view: "new varieties live long based on management. Because if you leave Kibuzi [a traditional variety] in the bush for one year it continues to yield, but the Kawanda varieties cannot last longer than one season if they are not cared for ... we prefer the old varieties because the news ones don't last long."[71] A male farmer in Nakaseke District corroborated these findings, commenting: "I will keep to [the traditional variety] because as you keep replacing, the farmer keeping [the traditional variety] will only be harvesting and avoiding labour costs and it takes you back in time ... I don't want to plan every year."[72]

A related issue concerns predictability and risk. Critics argue that promoters of GM technology assume that farmers are risk neutral and "seek to maximize their average profits, rather than minimize variance in their profits, or avoid years with low profits," when in fact smallholder farmers tend to be risk averse and prefer lower yields that are stable over time.[73] We set up a second experiment to test this hypothesis, asking farmers to choose between a variety with low but stable yields or one that produces higher short-term yields that decline over time. Again, farmers overwhelming preferred the former over the latter. [74] This tendency towards risk aversion came out in the focus groups as well. In Buikwe District, a farmer explained that he preferred the traditional variety because of its predictability: "I'm very sure that I'll get a bunch."[75] A male farmer in nearby Nakaseke District explained that predictable yields "make it possible to make long-term contracts with buyers. That's not possible with fluctuating yields."[76]

Farmers in our sample expressed other frustrations with the formal seed sector. One common grievance concerned issues of transparency and accountability. Female farmers in Buikwe District, for instance, pointed to the lack of "quality management," relaying that nurseries often trick farmers by giving them a variety inferior to the one for which they paid. In contrast, they can be assured of quality when buying from a neighbour because they see the performance of the parent plant with their own eyes.[77] Studies from elsewhere confirm the legitimacy of these concerns. One study in Kenya underlines the financial barriers farmers face in accessing the formal seed system, suggesting that the availability of improved varieties can be hampered via delivery through the formal seed system.[78] Another study in Ethiopia reveals farmer enthusiasm for improved varieties but suggests that diffusion and use were constrained by the formal seed system's ability to disseminate them.[79]

While the initial contact with nurseries might prove an obstacle to farmers, breeders advocate for reforming the informal system to ensure more transparency and accountability: "we would prefer a better system. For instance, when those tissue culture companies deliver material to the farmers, eventually those farmers distribute materials between themselves, but nobody knows the quality of those materials between the farmers. So there is no regulatory system in there. And a system that is not regulated, you don't know anything that goes on with that."[80]

One final important dynamic that emerged from our research project relates to how a GM version of matooke will impact male and female farmers differently. Previous research shows clearly that an enhanced production method like a new technological innovation will impact male and female farmers differently, largely because of differential control over key resources such as land, labour, and income, which ensure that the benefits associated with this new intervention accrue disproportionately to men.[81] Through our conversations with farmers we sought to identify the precise gendered impacts of a GM banana. For instance, a BBW-resistant banana will reduce the labour burden needed for disease prevention and control, while the associated higher yields will increase the labour required for harvesting. In the central and eastern regions of Uganda, where farms tend to be smaller and subsistence-oriented, farmers are more likely to intensify their use of unpaid family labour, particularly that of wives, to meet these additional labour demands. As such, the GM banana has the potential to exacerbate the workload for this heavily burdened and vulnerable demographic, and to entrench inequities between male-headed households better equipped to take advantage of these additional labour demands and female-headed households that are more likely to struggle to reap any benefits.[82]

CONCLUSION: IS NARO'S APPROACH TOO NARROW?

Matooke banana is, in many ways, an ideal candidate for genetic modification. It is basically infertile, very difficult to improve via conventional breeding, and is geographically confined to a single part of the world. It is also a crop that is eaten almost exclusively by poor people, and thus presents a unique opportunity to evaluate the potential for this new wave of GM orphan crops to improve the lives of smallholder farmers.

This chapter has compared the assumptions possessed by donors and breeders with perspectives gathered from a representative sample of Ugandan farmers, exposing the disconnect between the aspirations and understandings of those who direct the GM matooke program and those who are intended to benefit from it. This process reveals some potential flaws with the experimental program as it is currently conceived. First are complications that arise from knowledge politics – that is, the structures and incentives that have shaped the current experimental program. The heavy-handed involvement of development donors, especially the BMGF and USAID, has created a divide between the traits being prioritized by benefactors (biofortification) and those prioritized by farmers (disease resistance). Government breeders are caught in the middle and tasked with reconciling outsider wishes that are often out of touch with on-the-ground realities. Breeders are thus forced to navigate a perilous middle ground, where the funding for their work, their career progression, and oftentimes their actual jobs compel them to please donors ahead of pleasing farmers. NARO breeders are overworked and underpaid, and like most of us they respond to the structured incentives of their jobs.[83] The result is a breeding program that prioritizes the donors' view over that of smallholder farmers.

Second, while embracing genetic modification as a breeding strategy makes sense given the difficulties in improving a vegetatively propagated crop via conventional means, the heavy investment in producing improved varieties overwhelms other alternatives for enhancing banana productivity. When asked what they consider to be the most useful intervention for maximizing productivity in banana, farmers in our sample pointed to investment in extension services and education around improved agronomic practice. For example, farmers coveted greater access to tissue culture banana, which offers disease-free plantlets via micropropagation. But there is little interest from donors in expanding or subsidizing such programs: as one senior breeder put it "donors are looking for innovative research. Tissue culture is old. Tissue culture is not innovation."[84] Other forms of rapid multiplication techniques such as on-farm propagation and community nurseries have also been empirically validated as effective methods for improving access to enhanced planting materials, but these need to penetrate hard-to-reach rural areas in order to be effective.[85] Expanding and entrenching farmer-driven propagation mechanisms alongside investments in education campaigns and extension services is an empirically validated method for increasing farmer access to improved planting materials. But the restrictive focus on genetic modification crowds out these alternative interventions.

The final issue is one of context. Early results suggest that the GM banana is demonstrating both desired traits of disease resistance and biofortification. But in order for farmers to replicate these successes on their own farms they will need to mirror the intensive growing regime employed on experimental farms, including the application of manure, proper spacing, and consistent trimming, weeding, and de-matting. The increased financial, time, and labour demands that will accompany GM matooke will be easier for some farmers to manage than others: those with larger plots, those who are more commercially oriented, and those who are well-connected within farming associations and with extension agents are poised to benefit more than their less-endowed contemporaries. Women are especially at risk of being left out: female banana farmers expressed more hesitation around GM technology, as well as concerns that any gains in productivity enhancement would increase their labour burdens.[86] A final point about this enabling context relates to the delivery system: farmers expressed hesitation regarding the host variety into which the GM traits are being inserted and the cost of tissue culture plantlets, which could be a barrier of entry for poorer farmers with minimal land holding. While many of the breeders I have befriended over the years dismiss this slate of concerns as an "unnecessary controversy," I remain unconvinced that the project will succeed unless it is reconceptualized to take farmer preferences into account.[87] While the GM banana certainly holds promise, these questions around its fit within smallholder agricultural systems need to be addressed for it to achieve its worthwhile goals.

Conclusion

This book has surveyed twenty years of efforts to employ genetic modification as a tool to alleviate poverty and hunger in Africa. So what have we learned? First-generation GM technologies such as Bt cotton and Bt maize do not fit well within African smallholder farming systems. By and large, these are single-event products that require farmers to sacrifice acreage to refugia in order to mitigate resistance, with accompanying prohibitions on sharing seed and intercropping. First-generation GM crops were developed to succeed within the confines of industrial agriculture – large-scale, heavily capitalized, mechanized monoculture. Transplanting this technology to smallholder African farmers who operate in a completely distinct mode of production has failed. Expect this trend to continue into the future.

Second-generation GM crops offer a compelling redress to many of these shortcomings. They focus on crops and traits that matter to smallholder farmers and they are not bogged down by cumbersome intellectual property agreements. This book has uncovered the obstacles that scuttled the two pioneering precedents, insect-resistant maize and virus-resistant sweet potato, and raised questions about two of the most heralded GM orphan crops set for commercial release, WEMA maize and GM banana. The WEMA program undermined its promise by prioritizing Monsanto's insect-resistant Bt gene alongside the primary focus on drought tolerance, which will leave adopters facing the same punitive approach for mixing, recycling or sharing seed that was spurned by Bt maize growers in South Africa. The decision to emphasize hybrid seed and the question of preventing resistance present additional obstacles regarding WEMA's fit within smallholder farming systems. I am more bullish on the prospects of GM matooke, especially the experimental stream focused on developing varieties resistant to banana bacterial wilt (BBW). In this case a pernicious disease is ravaging the Uganda countryside, leaving farmers with desperately few options for management or control. And banana is a crop that is effectively sterile, which makes conventional breeding particularly difficult. For these reasons, I believe that the BBW-resistant variety deserves to be pursued.

My concern about all of these experimental programs is that they remain mired in some of the ruinous patterns that undermined previous agricultural development programs, including an agenda that is being imposed largely from the outside in and a misunderstanding of the farming systems in which these technologies are designed to function. Soon-to-be-released GMO 2.0 varieties are unlikely to succeed where their first-generation predecessors have failed unless these deficiencies are addressed.

THE YIELD GAP VERSUS
THE CONTEXT-METHODS-SCALE GAP

The entry point for this book – and the foundational imaginary for the debate around the potential for genetic modification to enhance African agricultural productivity more generally – is that of the yield gap, the measurable difference between potential yields achieved on experimental stations and actual yields realized on farmers' fields. This book has underlined the limitations of using the yield gap as the fulcrum for Africa's Gene Revolution: it prioritizes yield over other traits that are often more important to farmers, it assumes a linear association between yield maximization and poverty alleviation, and it privileges issues of technological enhancement over other social and political dynamics that play crucial roles in determining livelihoods. A political ecology of Africa's Gene Revolution reveals that any assessment of genetic modification's potential to address the yield gap needs to pay concomitant attention to three vital issues of context, methods, and scale. While the notion of a context-methods-scale gap is neither as pithy nor as catchy as that of a yield gap, these dynamics are integral to producing a systematic assessment of the potential for the Gene Revolution to alleviate poverty and hunger in Africa. Taken together, context, methods, and scale represent the missing link in evaluating whether and how GM technology can enhance smallholder farming systems.

The first underappreciated element relates to context. New technologies cannot be evaluated in isolation from the context in which they are introduced. This book has undertaken a comprehensive analysis of the fit between each GM technology and the farming system it is designed to enhance. The results presented here reveal weighty questions that remain underappreciated within the current enthusiasm for Africa's Gene Revolution, including its ability to sync with farmers' cultural preferences; the importance of concomitant and lasting corollary investment in credit, market access, and labour; and a disconnect between the traits being prioritized by donors and those preferred by farmers.

So how important is a supportive context to the success of GM technology? The instances where first-generation GM crops succeeded in boosting yields and livelihoods were those in which the context was transformed to support their adoption: in the Makhathini Flats, where accompanying credit and restrictive buying arrangement buoyed sky-high early adoption rates, and in Burkina Faso, where a vertically integrated monopoly was able to guarantee competitive returns. When these supportive contexts disappeared, so too did the technological benefits. The experience of Bt maize in South Africa reveals a similar pattern. The economic benefits to large-scale producers of yellow maize are convincing, but both insect-resistant and herbicide-tolerant varieties clash with the growing regimes of smallholders, whose limited land holdings, close proximity to neighbouring farmers, and predilection for sharing and recycling seed contravene the trait segregation practices that are essential to the technology's success. Sudan offers the most promising setting for Bt cotton, largely because its government was able to institute structural incentives and a centralized industry necessary to take advantage of economies of scale. GM cotton and GM maize offer a more remunerative form of agricultural production for the industrial production systems within which they were designed to succeed. But cotton and maize are produced in vastly different contexts in Africa. These first-generation GM technologies cannot succeed unless the context is modified to support these technological prerequisites.

GM versions of orphan crops seem to represent a better fit between technology and context. But careful attention to context reveals an important distinction based on the growing requirements of the crop itself. Efforts to breed insect-resistance for maize via both the IRMA and the WEMA program fell into some of the same traps detailed above, including onerous requirements around intellectual property rights, compliance with biosafety requirements, and restrictive growing regimes. This stems both from maize's reproductive biology as an OPV and its primacy as a global crop for both food and feed. Vegetatively propagated crops such as potato, banana, cassava, and sweet potato represent more logical candidates for genetic modification because they are notoriously difficult to improve via conventional means. Current versions such as cassava that is resistant to cassava brown streak and cassava mosaic disease, and matooke that is resistant to BBW, address a constraint that is meaningful to smallholder farmers; plus, the crops' self-replication makes the kind of rigid intellectual property and growing restrictions associated with cotton and maize untenable.

The other important contextual element relating to the crop itself concerns the choice of host variety. Case studies from this volume underscore that successful GM breeding hinges on the interplay between GM trait and host cultivar. Complications surrounding the introgression of Monsanto's Bt construct into the high-quality Burkinabè cotton doomed commercialization efforts in that case. Both IRMA's initial attempt to create GM OPVs and the GM sweet potato's insufficient commitment to developing clones suitable to each of Kenya's distinct agro-ecological zones underline the importance of choosing an appropriate host. So too does the case of matooke banana, with which farmers repeatedly emphasize their preference for traditional varieties based on value-based characteristics such as taste, colour, and texture.

Taken together, these insights remind us that any new technology is only as good as the context into which it is inserted. Seed-based technological improvements like genetic modification will only prove advantageous for smallholder growers if an enabling environment accompanies them. Both IRMA and the GM sweet potato project presumed that an efficient and effective seed system was in place that would facilitate farmer access to these technologies once they were ready for commercialization, but neither invested in these crucial arenas of distribution, extension, or education. Similarly, both WEMA and the GM matooke program hinge largely on their ability to create dissemination mechanisms congruent with existing informal seed systems. The take-home message here is that investments in seed-based technology must be accompanied by concomitant investments to ensure that supportive structural factors are in place in terms of credit, markets, extension, outreach, and education. Melinda Smale puts it nicely: "For seed-based technical change to succeed, it must be accompanied by strong rural institutions that enable farmers to understand the crop they are adopting, learn from growing it and be able to rely on input and output markets from one year to the next. Seed is a minor contributor in the quest for poverty reduction, although it can, and has, served as a catalyst. Investments in rural development should therefore accompany investments in biotechnology."[1] This relentless focus on context showcases why supposed "win-win" partnerships between the private sector, the state, and philanthropic donors have not translated into positive results on the ground.

The second missing piece here relates to methods. The science of genetic modification is moving faster than the social science evaluation of the technology's capacity to achieve its lofty goals. The vast majority of studies that assess the potential for GM crops to improve agricultural development rely on econometric models or large-scale surveys. These top-down methods obscure many of the context-specific issues identified above, which play a critical role in determining whether farmers actually adopt these new varieties. Some of the methodological limitations here are obvious, such as the high proportion of assessment studies undertaken by individuals or institutions that are also responsible for the dissemination of these biotechnologies. This lack of autonomous, independent assessments has cropped up throughout the book, including with INERA documenting the triumph of Bt cotton in Burkina Faso, CIMMYT evaluating the future implications of WEMA maize in Kenya, and NARO evaluating the farm-level impacts of GM matooke in Uganda.

Some methodological limitations are less apparent. Anthropologist Glenn Stone has laid bare the subtler deficiencies that plague much of the existing research base, including the absence of counterfactuals, selection bias that creates samples made up primarily of early adopters who are not representative of broader populations, and cultivation bias by farmers that leads them to prioritize GM varieties relative to their conventional counterparts.[2] We have encountered examples of each of these in these pages. Existing accounts that showcase the success of Bt cotton in Burkina Faso were based on comparisons with conventional yields on nearby refuges that served as buffers to mitigate gene flow; any yields produced here were accidental and artificially low.[3] My own research on GM matooke in Uganda exposes the dangers of selection bias, in which early adopters disproportionately belong to advantageous groups such as those with larger land holdings and better access to extension agents.[4] Issues of cultivation bias seem likely in the case of both WEMA maize and GM banana, where the vessel in which the GM trait is embedded – hybrids in the case of WEMA and tissue culture in the case of matooke – will require more inputs from the growers who plant them.

Research programs need to redress three key shortcomings that become apparent via the lens of political ecology. First, we need longer-term studies of the impact of genetic modification on African farming systems. The recklessness of short-term studies of technological triumph has been exposed in the case of Bt cotton in the Makhathini Flats and Burkina Faso. We need longitudinal studies that evaluate whether short-term gains buoyed by selection bias and cultivation bias can be sustained over the long-term. Second, we need research that exposes the differentiated impacts that accompany the introduction of any yield-enhancing technology, especially as it relates to gender. Scholars agree that underlying social inequities between men and women will play a crucial role in shaping outcomes associated with the introduction of GM crops, yet there is scant empirical evidence assessing the impact of these technologies on gendered dynamics. The little research that does exist is largely contradictory: studies in Uganda predict that GM banana will intensify the time and labour burden for women farmers,[5] while research on Bt cotton in Burkina Faso and Bt maize in South Africa found that GM technology lessened women's work in terms of both land preparation and weeding.[6]

The third redress relates to farmer decision-making. Econometric models and large-scale surveys tend to portray farmers as monolithic actors who will choose increased yields and increased profits every time. But there is a vast array of scholarship showing how complex and multifaceted farmer decision-making really is, with innumerable examples of farmers forgoing increases in yield in favour of other valuable traits such as early maturation, favourable processing characteristics such as poundability and storability, and, especially, consumption characteristics such as taste, colour, texture, and familiarity.[7] Similarly, the limited research on farmer adoption of GM crops in Africa underestimates farmer risk aversion. Studies have shown time and again that smallholder farmers tend to privilege livelihood security over yield maximization, a dynamic that is consistently underappreciated in predictive modelling built upon principles of modernity and rationality.[8] More broadly, the notion of technology adoption itself needs to be problematized, with more grounded accounts of farmer decision-making showing that uptake of new technologies is much messier than previously understood, marked often by a transition to and integration of new technologies into existing farming systems as opposed to a one-to-one replacement. These methodological biases help to explain why genetic modification has been favoured over other technological possibilities and why certain versions of genetic modification have been favoured over others.[9]

Scale is the third element of this missing link. Much of the enthusiasm that surrounds GM crops is premised on their ability to impact small-scale agriculture. Florence Wambugu, one of the continent's most celebrated champions of biotech who spearheaded the GM sweet potato introduced in chapter 5, extolled the potential of genetic modification based on its "packaged technology in the seed, which ensures technology benefits without changing local cultural practices."[10] Elsewhere, Wambugu argues that seed-based technology is "more easy to adopt for smallholder African farmers, who might have little education, but have indigenous knowledge and experience on how to handle seed."[11] This emphasis on seed-based technology weaves its way through the pro-biotech messaging. In an interview with Wambugu's organization, Africa Harvest, which is trying to pioneer the world's first GM biofortified sorghum, her colleague explained that genetic modification's potential for smallholders is due to the technology in the seed, which guarantees that "the farmer's practices do not change."[12] Others similarly argue that seed-based technology is particularly amenable to scaling up, since it is "equally applicable on small or large farms."[13]

This book challenges these linear interpretations of the relationship between genetic modification and scale in the African context. Changes at the genomic scale have impacts at all levels. New technologies precipitate shifting institutional arrangements and create new incentives for farmers. The narrow view that seed-based technologies are neutral to scale ignores the interactions between technology and context that shape farm-level realities. Access to information, access to complementary inputs, credit, transport, intra-household realities – all of these vary in relation to farm size.[14]

This preoccupation with scaling up reveals the importance of agricultural modernization discourses within genetic modification's rise to prominence. At its core, the Gene Revolution is about transforming African smallholder agriculture into a mode of industrial production: large-scale, monocropped, heavily capitalized, and input-driven. Development donors are unapologetic on this, arguing that scaling up agricultural production is a prerequisite for transforming African agriculture into commercial production.[15] Critics have attacked this preoccupation with scale, arguing that agricultural modernization represents an attack on small-scale agriculture, as it works to consolidate autonomous plots into large-scale commercial ventures that are more easily incorporated into global value chains, thereby displacing small-scale producers and transforming them into wage labourers.[16]

The evidence presented here suggests that the blind spot of the Gene Revolution is its inability to downscale – that is, the capacity to adapt biotechnologies to suit local practice. The major obstacle here is the linear technology pipeline, a mode of addressing urgent development needs by creating new products to meet demand that is presumed, rather than confirmed.[17] The experimental regime of the Gene Revolution is beholden to the model of P3s that dominates the current landscape of agricultural development. This regime's characteristics have been documented repeatedly throughout these pages: a trial-driven, top-down system adherent to results-based frameworks, rigid delivery schedules, short-term targets, onerous requirements for transparency and accountability, and above all, providing value for money.[18] Methodological propensities for modelling and surveys being used as tools to aggregate farm-level challenges serve to erase the local scale in favour of higher-magnitude conclusions, leading to interventions that oversimplify multiscalar complexities and clash with the lived realities of smallholder farmers.[19] The restrictions embedded in GM technology – especially those related to predictable growing patterns, prescribed spacing, limitations on intercropping, and the exclusion of recycled and replanted seed – further compound the inelasticity of these programs and put even more onus on the directives and priorities of the technology developer relative to the intended beneficiary.[20] Current Gene Revolution thinking oversimplifies the nested, overlapping systems that shape African farmer systems at multiple scales.[21] Breeding programs that seek to engage with these cross-scale dynamics, rather than occlude them, are more likely to reconcile development planning and farm-level experience.

STORIES, POWER, AND POLITICS

These three inextricable issues of context, methods, and scale combine in different ways to offer contrasting assessments of the potential of genetic modification to enhance smallholder agricultural systems in Africa. The resultant narratives of technological potential evoke moral, political, and cultural ideals and shape meaning in important ways. Individuals and institutions invested in particular outcomes use their financial, political, and legal power to create and perpetuate narratives that serve their interests. At its core this process of technological storytelling is one of representation, in which carefully constructed narratives are crafted and upheld by powerful actors seeking to advance their particular political agendas.

This book has attempted to unravel the various components of technological storytelling and has exposed how representations of context, methods, and scale have been used to advance certain framings of Africa's Gene Revolution. The prevailing narrative across the continent is one of technological inevitability. Embedded within an unquestioned exaltation of progress, this narrative subsumes genetic modification within a teleological view of economic growth, imbuing it with a sense of destiny: genetic modification is the future. Genetic modification is venerated as science, while any opposition is dismissed as politics. Any country choosing to resist this technological advance risks being left behind.[22]

This framing of technological inevitability is apparent in the case of Bt cotton, in which depictions of the initial successes of Bt cotton in both South Africa and Burkina Faso long outlasted the less enthusiastic realities encountered by farmers on the ground. It is equally pronounced in the asymmetrical power relations that underlie the partnership approach – particularly the emphasis on P3s within GMO 2.0 – in which the technical aspirations of donors lead to experimental programs that put the solution before the problem, as evidenced by the fixation on transplanting existing technology such as Monsanto's coat protein conferring virus resistance (in the case of sweet potato) and the Cry proteins conferring insect resistance (in the case of IRMA and WEMA). This narrative of technological inevitability is further buoyed by the compartmentalized approach employed by breeders committed to a "one constraint, one trait" approach, such as in the case of virus-resistant sweet potato or biofortified banana, both of which failed to recognize the complex and interrelated nature of growing constraints for smallholders. The result has been experimental programs that are capable only of generating technical solutions to political problems.

On the other side of the divide is a complex of NGOs and activists willing to seize upon any crumb of doubt or uncertainty in order to attack GM crops as a nefarious plot of intensifying corporate control.

Glenn Stone exposes the self-affirming nature of these hardliners, who reject GM crops as an attempt by multinational corporations to subsume Africa within their control.[23] The anti-GMO narrative propagated by opponents is one of David versus Goliath, with small-scale farmers positioned as heroes withstanding corporate incursion. In attacking Africa's Gene Revolution, these anti-GMO framings fail to acknowledge the substantive differences between first- and second-generation GM crops as they pertain to intellectual property rights, crop selection, or target traits. Some denunciations are even more extreme. In Ghana, the primary conglomerate of anti-GMO campaigners produced a series of banners and pamphlets implying a relationship between GM crops and the Ebola virus via the slogan "GMO/Ebola Out of Africa." ActionAid Uganda was formally reprimanded by its parent organization for indicating that GM foods cause cancer, at which point the organization admitted to misleading the Ugandan public and formally withdrew these claims.[24] Such overstated claims are common within the debate over genetic modification in Africa, where suggestions of impaired fertility and carcinogenic properties are rife. Pro-GMO campaigners have vociferously attacked this campaign of "disinformation ... led and financed mostly by individuals from well-fed countries who do not need the technology themselves," whom they argue are complicit in unnecessary deaths that GM crops could have helped to avoid.[25]

As Stone explains, these sparring sides produce conflicting stories that each reinforce their own political leanings while avoiding any systematic engagement with actual evidence.[26] One example from my own work is relevant here. Research showed that declines in cotton quality were the main factor that precipitated Burkina Faso's reversal on GM cotton, which challenged the dominant representation of technological inevitability in Burkina Faso. The response from proponents was not to dispute the results but rather to massage them so that the overarching narrative remained intact: supporters portrayed Burkina Faso's decision to suspend Bt cotton as "temporary" and an "exaggeration," seizing upon the notion that the introgression issue related to "basic crop breeding, unrelated to the genetically engineered traits," which affirmed their confidence that such minor hurdles would eventually be overcome in time.[27] At the other end of the spectrum, anti-GMO activists sought to portray Burkina Faso's context-specific plight as a harbinger for change across the continent, one "that could help decide the future of GM crops in Africa."[28] Both sides spun stories regarding this research to further their own campaigns.

The major battleground for these competing stories is the realm of governance, with each side trying to establish regulatory regimes that are as permissive or prohibitive as possible. Anti-GMO NGOs and activists notched early successes with the African Model Law and the much-publicized Zambian rejection of GM food aid. In response, BMGF and USAID created intermediary organizations such as the Program for Biosafety Systems (PBS) and the African Biosafety Network of Expertise (ABNE), which crafted multiplatform outreach campaigns combining radio programming, "seeing is believing" tours, sensitization programs, and lobbying of key policy makers. Resisters such as Zimbabwe and Tanzania, the as-yet-to-be-dismantled Kenyan cabinet ban, and Uganda's surprisingly restrictive new biosafety bill offer further evidence that the anti-GMO camp continues to shape policy outcomes. But the "anti" side simply cannot match the financial resources and political clout of their opponents. Ghana's high court decision upholding its biosafety bill, alongside the proliferation of emerging adopters including Nigeria, Malawi, Ethiopia, Mozambique, Swaziland, and Cameroon, suggests that the narrative of technological inevitability is steadily gaining ground. This blueprint of a permissive and harmonized framework being driven largely by outside donors remains the most common regulatory approach across the continent.

It is in these manifestations of knowledge politics that the depth and scale of the powerful interests underpinning Africa's Gene Revolution are fully exposed. Decisions around what crop and trait to modify, what gets measured in evaluating outcomes, and how to regulate GM crops are the arenas in which competing narratives around Africa's Gene Revolution become entrenched. It is in these spaces of contestation that power and politics combine to produce stories that reinforce established positions: a narrative of technological inevitability abuts a narrative of corporate control. This polarizing process of technological storytelling serves to deepen the gap between opponents on either side of the debate, undermining the possibility of constructive engagement and overpowering the more fragmented stories recounted by the smallholder farmers who are the technology's intended beneficiaries.

THE WAY FORWARD

Books like this one are often dismissed as being all about critique; after all, it is easier to attack the Gene Revolution than suggest alternative strategies for improving agricultural development in Africa. So how best to move forward? Let's begin by identifying the wrong way forward. Current debates over genetic modification have become so polarized that they amount to little more than academic shouting matches. The intractable "pro" versus "anti" divide leaves no room for constructive conversation. Moving forward means recognizing that answers do not lie at one extreme pole or the other but somewhere in the middle. We need to create a conversation around Africa's Gene Revolution that recognizes broader global issues and precedents but examines these issues primarily in the context of African agricultural systems. Nearly all of the research, policy, and outreach being undertaken on the continent is being funded by one of these extremes: either by development donors such as BMGF or USAID, who are pushing for the expansion of genetic modification, or by civil society groups committed to opposing genetic modification in all its forms. We need neutral platforms funded by institutions with no self-interest in the outcome. The models employed by the National Academy of Sciences in the United States or the European Food Safety Authority offer useful templates.[29] The United Nations' FAO or the African Union would be well positioned to open up the conversation to both sides while adhering to evidence-based practice, in order to better navigate this messy middle ground and empower African farmers to make informed decisions around these new technological possibilities. We need evidence-based encounters that shift the conversation away from being for or against GM crops, towards a more nuanced exchange that zeroes in on the key issues of context, methods, and scale identified above.

This revamped conversation needs to be data driven. In that spirit, I propose an evaluative framework that can be used to assess the viability of GM crops – or any agricultural biotechnology more broadly – based on the three core issues of context, methods, and scale. Such a framework would be anchored by the following criteria:

Context: Do target crops and traits match up with farmer needs? Do the growing requirements for the GM crop sync with existing smallholder systems, and if not, how are differing requirements in terms of monocropping, extra inputs, refugia, and restrictions around replanting going to be implemented? Is there an accompanying commitment to creating an enabling environment in the form of credit, market, education, and extension?

Methods: Is there credible, independent research showing gains for smallholder farmers? How will the impacts associated with this new technology be differentiated according to gender, wealth, age, and other categories of power?

Scale: Has farmer input been meaningfully incorporated throughout the project's life cycle? What assumptions, simplifications, or consolidations about local farming systems are embedded within the project?

I am not suggesting that this framework is the only mechanism for assessing whether these key dynamics have been addressed. Other formulas have been proposed that use different questions to ascertain similar information. One of the most promising is Jack Kloppenburg's notion of seed sovereignty, which outlines four components – the right to save and replant seed, the right to share seed, the right to use seed to breed new varieties, and the right to participate in shaping new policies for seed. Applying these criteria to Africa's Gene Revolution reveals that first-generation GM crops are too limiting and restrictive to receive endorsement, but that second-generation GM crops go much further in satisfying the first three (though there is still a long way to go for farmers to be fully participating in shaping policies around seed). A second promising alternative is the STEPS Centre's 3D approach to considering technological pathways, which encompasses a focus on directionality of change, distribution of benefits, and diversity of impacts.[30] We need not quibble about how these criteria are phrased, but we do need to ensure that they are integrated as fundamental principles within the evaluation of biotechnological interventions.

From evaluation we need to move towards improved execution. The existing body of evidence is unequivocal: the Gene Revolution's technology pipeline is broken. This inelastic model needs to be reconceptualized and replaced with an iterative experimental structure that is more nimble, flexible, and responsive. This new approach needs to be less ambitious in terms of timeline and expectations, supported by longer-term grants with targeted funding for farmer experimentation and adjustment to facilitate co-learning, and increased investment in rigorous social science to evaluate whether proposed experimental solutions make sense for farmers on the ground. P3s such as WEMA and the GM matooke program need to be reformed to redress asymmetrical power relations, propensitites for technological interventions, and assumptions around farmer rationality, which lead to interventions that place donor priorities and perspectives above those of the farmers themselves. The Gene

Revolution model needs to do away with the current cursory nod to consultation, where farmer input serves as a bookend sought once at the beginning of the project in order to justify the chosen scope, and once again at the end to confirm the chosen strategy. We need to reimagine technology transfer as technology co-development, in which so-called participatory processes are opened up and decentralized, and lopsided power relations are rebalanced to ensure that farmers are integrated in meaningful ways throughout all project phases.[31] Yes, this new way of doing things will be more expensive, and yes, it will take more time. But better to overhaul the current system than continue to repeat these same mistakes.

Scholars have uncovered credible examples of what a radically overhauled breeding program might look like, examples in which the seed-based technology is decentred and revolves instead around the farmer as expert. Co-developing technology requires abandoning the model of transplanting existing first-generation GM crops to Africa, and ensuring instead that Africans have the resources – in terms of finance, capacity, infrastructure, and political power – to identify and develop these technologies for themselves. Whitfield reviews a host of programs that have been reconstituted as agricultural encounters between researchers and farmers, including conservation agriculture, integrated soil fertility management, and agroforestry, each of which seeks to create technologies that function as a loose assemblage of agronomic principles designed to be tailored at the farm level.[32]

Another strand of scholarship reveals how these principles of coevolving "social, technological and environmental systems over time" can be operationalized in practice.[33] Alternately known as the innovations system approach, agricultural innovations systems, farming research systems, and the STEPS pathways approach, this perspective involves a commitment to understanding knowledge development and application as two halves of the same whole. The debate over genetic modification would benefit from the call by Giller et al. for a new "systems agronomy," in which agricultural research is transformed into a place-based science that would provide an "empirically grounded, adaptive approach" to sustainable intensification, one that "does not merely focus on production and environment, but calls attention to social acceptability and economic viability."[34] Ethiopia, for instance, was able to triple its maize yields within ten years by crafting a homegrown platform by investing holistically in successful agricultural systems, which included extension services and post-harvest storage alongside improved seeds and inputs.[35] Some of the most successful breeding programs on the continent have embraced these commitments to place-based knowledge exchanges bolstered by farmer agency and capability.[36] One such example was detailed in chapter 4 – that of the Southern African Drought and Low Soil Fertility (SADLF) project, which was anchored in low-input and high-stress environments encountered by smallholder maize farmers and bred improved versions of OPVs that responded to these and other farmer priorities including poundability, favourable taste, and minimal loss during the milling process. Another innovative and effective program is the plant clinics run by the Centre for Agriculture and Bioscience International (CABI), in which mobile clinics are set up at rural markets to allow trained experts to support farmers in diagnosing and treating pests or diseases.[37] This decentralized model in which experts come to the farmers is rooted in place-specific queries, and empowers the farmer to set the agenda and prioritize farm-level concerns. A systems-agronomy approach could help move beyond Bill Gates's "big bets" on the yield gap, as laid out in the introduction, replacing it with an emphasis on "best fits" between technology and context.[38]

Agro-ecology offers another promising method for empowering farmer voices within technology development programs and creating interventions that are embedded in a particular social, ecological, and cultural context.[39] Defined simultaneously as a movement, a science, and a practice, agro-ecology refers to the "application of ecological concepts and principles to the design and management of sustainable agricultural ecosystems."[40] Agro-ecology has achieved mainstream status within agricultural development programming, having been endorsed by both the International Assessment of Agricultural Knowledge, Science and Technology (IAASTD) and the former UN special rapporteur on the right to food, who argued that agro-ecology increased productivity, reduced poverty, enhanced nutritional diversity, and augmented resilience.[41]

A key emphasis within agro-ecological approaches is transforming farmers into co-designers in knowledge production by "restoring their roles as pilots of farming systems."[42] There exists a flurry of promising examples for how such principles could be operationalized in practice. Successful templates include the Seven Ravens Permaculture Academy, which facilitated co-living and co-learning between farmers and experts in Kenya;[43] the Soils, Food and Healthy Communities project in Malawi, which organized farmer research teams tasked with experimenting on improved soil fertility practices and shared this insight via farmer-to-farmer teaching;[44] the Ikulwe Project, which implemented a decentralized and socially equitable model that brought farmer needs to the forefront throughout the entire research life-cycle into participatory crop breeding in Uganda;[45] and the Kotoba Sustainable Livelihoods Project, which used farmer-owned demonstration sites and self-help groups to leverage farmer capacity and increase farmer sharing.[46] Another promising example is that of farmer research networks, which connect large, representative groups of farmers who take ownership over all stages of experimentation in order to achieve an agro-ecological intensification rooted in principles of "mutuality, reciprocity, beneficiary ownership and local agency."[47] Each of these interventions commits to social learning and empowering farmers to experiment with new information and innovation. In so doing, they disrupt the Gene Revolution's locked-in theory of change, which remains committed to market-led technology for development, replacing it with a more open-ended approach focused on galvanizing existing informal seed systems.[48] Agro-ecology serves as a counterpoint to the dominant model embraced by Africa's Gene Revolution, in which farmer knowledge and practice are considered less important than the technological prowess of the new intervention, shifting the focus instead to programs that privilege farmer skill, competency, and agency.[49]

The alternatives surveyed here expose the core limitation of Africa's Gene Revolution: its model is based on tweaking technological inputs to bring about major change, ignoring the need to radically transform the broader structures that constrain African smallholder producers. Changing a plant's genome is an easier task than changing an entire system. There are fewer variables to control, fewer actors to consult, and fewer barriers in the form of policy, politics, and people. But the core message of this book is that any benefits associated with genome enhancement cannot be realized without systemic change.[50] To achieve this, we need to move from genetic engineering towards agro-ecological engineering – that is, investment in technological improvement that is not just focused on the genomic scale but instead tackles the multiplicity of factors that limit smallholder production on the continent.[51] An emphasis on systems-based intensification moves beyond the reductionist emphasis on genetic improvement and the narrow focus on redressing yield gaps towards investments in agro-ecological systems, which may or may not require technological improvement. This approach rejects the premise that agricultural technology can be standardized and scaleable, starting instead with location-specific conditions to develop interventions that make farming systems more productive and resource efficient. Lasting change will only be realized by radically rethinking African agricultural systems, not just the plant's genome.

FINAL THOUGHTS: AGRICULTURAL TECHNOLOGY FOR DEVELOPMENT

My final point is that we must look backwards in order to move forwards in the debate over Africa's Gene Revolution. The history of agricultural development in Africa offers a litany of precedents showing what does *not* work: interventions imposed upon African farmers by experts with limited understanding of local farming systems, who champion science and technology as the sole means of enhancing productivity, and employ a compartmentalized approach that views ecological, social, and political issues as discrete and independent. I am worried that Africa's Gene Revolution replicates many of these same patterns: by relying on technologies that are supply-driven rather than demand-driven, by setting priorities outside rather than inside the continent, and by putting too much emphasis on the technology itself as opposed to the enabling structure necessary for the technology to succeed. The most common refrain that comes up in my conversation with policy makers is that the debate over GM crops needs to be grounded in science, not politics. But taking a long view on African agricultural development makes clear that there is no separating out these two realms. All science is political. Deciding which technologies to fund, which crops to modify, and which traits to prioritize are value-based decisions that reflect the priorities of the powerful benefactors of this paradigm over those of the smallholder farmers it is designed to help.

These conversations are only going to become more complex. A new suite of gene editing biotechnologies – including the much-hyped clustered regularly interspaced short palindromic repeats (CRISPR) – present tantalizing possibilities for agricultural development. Already, advocates are heralding the potential for gene editing technologies to revolutionize African agriculture by offering an inexpensive, user-friendly, and accelerated mechanism for altering targeted genes of interest. Gene editing further represents a more palatable option to African farmers, consumers, and governments because it does not involve the transfer of foreign DNA, leaving open the possibility that varieties modified via CRISPR may be able to bypass the contentious debates around biosafety regulations surveyed in chapter 1.[52] Already, gene editing is being used to make plants more resistant to pathogens in rice and banana,[53] and eliminate detrimental traits in cassava.[54] CRISPR's ease of use and low cost may make genome editing a viable option for traits and crops that have been ignored by years of investment and innovation.

This hoopla of technological optimism should sound familiar. My bet is that the cycle of technological inevitability will start all over again. Cutting-edge biotechnologies using gene-editing techniques will be developed initially for profit-making purposes. Soon after, donors will seek to explore their prospective applications for humanitarian purposes. Eventually the conversation will turn to Africa, where all sorts of novel applications will be advanced as solutions to the continent's most intractable problems.

We need to treat this optimistic refrain around technological transplantation with suspicion. It boasts a dreadful track record within African agricultural development. We can now look back over two centuries of colonial and state-led agricultural development efforts, nearly all of which started with the premise that African farming systems are backwards, unproductive, and in urgent need of transformation.[55] Joseph Hodge has traced this commitment to improving African agricultural systems back to the earliest colonial agricultural ventures.[56] Fuelled by expert knowledge, imbued with an unbridled faith in science and technology, and rooted in a desire to manage the proliferation of multiple crises in the late colonial world, this was the site of the original agricultural revolution, which was heralded by colonial officers as a means of transforming African agrarian structures to incorporate the transmission of modern inputs and mechanized equipment.[57] This colonial agricultural revolution was propelled by a twin commitment to the "moral and material advancement of colonial people" alongside the opening up of the tropics to the benefits of European trade. Its impact was to recast "social and economic problems as technical ones that could be fixed by rational planning and expert knowledge."[58]

Again, this language should sound familiar. Africa's Gene Revolution epitomizes the modernist approach to agricultural development. The framing of both the problem and the solution echoes centuries of top-down, technocratic development planning. Africa's Gene Revolution is justified based on a conflation of crises that include stagnating yields, population growth, and climate change, its interventions packaged as an updated configuration of science, nature, and capital designed to modernize African farmers, while its impacts "represents a technical quick fix to food insecurity, reducing a difficult multi-dimensional and value-laden problem to a rational and technical linear inevitability – the political subsumed by the biological."[59] Examining the litany of historical precedents reveals that those productivity enhancement programs that paid close attention to place and context and existing farming systems were more successful than those that did not. And yet this remains one of the Gene Revolution's most glaring blind spots.[60] Aspirational commitments to participation and partnerships have proven insufficient as mechanisms for redressing these trends. The same tropes keep repeating themselves.

The legacy of agricultural modernization is most apparent in the attitude of the experts who champion Africa's Gene Revolution. I gleaned the greatest insights into the logic underpinning Africa's Gene Revolution from over 125 interviews undertaken over a period of ten years. Development donors are wedded to the ideal of problem solving, favouring interventions that are unconventional and aspirational. But rural poverty in Africa is not a machine with a widget that needs replacing or software whose code requires updating. Many solutions have cropped up in this book that offer empirically verified interventions for mitigating or managing urgent problems, including greater investment in OPV maize, mother-baby trials, credit bundles that include enhanced inputs, gender-specific education campaigns, farmer-to-farmer visits, tissue culture, and dietary diversification. What would be the result if half a billion dollars had been invested in this gamut of promising alternatives? We will never know. A host of empirically validated solutions remain mired in the backdrop of Africa's Gene Revolution because they do not mesh with donors' current preoccupation with disrupting existing systems via transformational innovation.

Breeders are equally entrenched in their own way of thinking, consumed with getting the science of Africa's Gene Revolution right.[61] The breeders I have spoken with over the years are compassionate and talented scientists who work tirelessly to develop technical improvements for smallholder farmers. But their understanding of their role as agronomists is largely restricted to their work on the experimental station. Many view farmer consultation as a box that needs to be checked in order to continue to receive the grants needed to fund their lab-based research. They remain wary of involving farmers too deeply in the breeding process: "in the final stages of testing the technology, when we have something in hand, that's when we involve the farmers."[62] Others resent technology co-development as an attempt to undermine their own expertise: "we think the farmers, even politicians, should trust the scientists, that we will do something that is safe for them."[63] The breeders who are delivering the science of Africa's Gene Revolution need to be accompanied by a cohort of social scientists from economics, anthropology, sociology, geography, gender studies, and development studies, who can work alongside them to transform the experimental model of the Gene Revolution into a "situated, place-based science" that can more effectively integrate farmers as co-learners.[64] Otherwise, breeders risk perpetuating experimental programs that ignore politics, powerful interests, and people, and, in so doing, may end up undermining the very science they are so committed to advancing.

There is a saying in Luganda: "in something good there is always something bad" (*mubuli kirungi, murimu ekibi*). Every new technology has benefits and drawbacks. This book has sifted through both primary and secondary data in order to highlight the implications of genetic modification technology for smallholder African farmers. Can GM crops improve agricultural productivity in Africa? Only time will tell. But more needs to be done to ensure that Africans are able to make these decisions for themselves. Donors, breeders, and critics need to present an honest and evidence-based assessment of these technological possibilities and then allow farmers to weigh the associated costs and benefits. We need to move beyond flawed models of technology transfer and invest instead in more iterative models of technology co-development that enable farmers to shape each and every stage of these experimental programs. Until then, Africa's Gene Revolution seems poised to repeat the mistakes of the past.

Notes

INTRODUCTION

1 Global Harvest Initiative, *2012 Gap Report.*
2 FAO, FAOSTAT; Pretty, Toulmin, and Williams, "Sustainable Intensification."
3 IFAD, *Rural Development Report, 2016.*
4 Delve et al., "Agricultural Productivity."
5 AGRA, *African Agriculture Status Report.*
6 Licker et al., "Mind the Gap," 769–82.
7 BMGF, "Our Big Bet."
8 I choose to use the term genetic modification throughout the book. While I recognize that a term like genetic engineering (GE) is more precise, and that all definitions are being challenged by emerging technologies like gene editing that make these categories even more difficult to define, I have chosen genetic modification because this is the term that resonates most soundly in academic, policy, and public debate.
9 Anthony and Ferroni, "Agricultural Biotechnology."
10 World Bank, *World Development Report 2008*; Ruane et al., *Biotechnologies at Work For Smallholders.*
11 BMGF, "GMOs Will End Starvation in Africa."
12 Lieberman and Gray, "GMOs and the Developing World."
13 AGRA, *African Agriculture Status Report*, 22.
14 Blein et al., *Agriculture in Africa.*
15 BMGF, "How We Work."
16 UN DESA Population Division, *World Population Prospects.*
17 FAO, *Smallholders and Family Farmers.*
18 McIntyre et al., *Agriculture at a Crossroads.*

19 Blein et al., *Agriculture in Africa*. See also Lowder, Skoet, and Raney, "The Number, Size, and Distribution of Farms."

20 Hazell et al., *The Future of Small Farms*. See also Netting, *Smallholders, Householders.*

21 FAO, *Sustainability Assessment of Food and Agricultural*; HLPE, *Investing in Smallholder Agriculture*, 10.

22 Sheridan, "An Irrigation Intake Is like a Uterus."

23 Green, "Production Systems in Pre-Colonial Africa"; Gareth, "Resources, Techniques, and Strategies South of the Sahara."

24 Friedmann and McMichael, "Agriculture and the State System"; Friedmann, "Feeding the Empire."

25 Badiane and Makombe, *The Theory and Practice of Agriculture.*

26 NEPAD, "Comprehensive Africa Agriculture Development Programme."

27 Blein et al., *Agriculture in Africa*, 83.

28 Ibid., 3.

29 World Bank, *World Development Report 2008*, 232.

30 Grow Africa. *Partnering for Agricultural Transformation*, 1.

31 New Alliance, "About."

32 Diao, Headey, and Johnson, "Toward a Green Revolution in Africa"; AGRA, *Strategy Overview for 2017–2021.*

33 Dalrymple, *Development and Spread of High-Yielding Varieties.*

34 Evenson and Gollin, "Assessing the Impact of the Green Revolution."

35 FAO, *The State of Food and Agriculture 2003–2004.*

36 Qaim, *Genetically Modified Crops and Agricultural Development*, 30.

37 Khush, "Challenges for Meeting the Global Food and Nutrient Needs in the New Millennium"; Khush, "Rice Breeding."

38 Boyce, *The Philippines*; Pingali, "Green Revolution."

39 Cleaver, "The Contradictions of the Green Revolution," 177; Pingali, "Green Revolution"; Sobha, "Green Revolution."

40 Singh, "Environmental Consequences of Agricultural Development."

41 AGRA, "Our Story."

42 AGRA, *Strategy Overview for 2017–2021.*

43 AGRA, "Our Story."

44 Interview with Informant 77, Nairobi, 14 May 2015.

45 Geggenbach et al., "Limits of the Green Revolution for Africa."

46 AGRA, "AGRA Focus"; AGRA, "Program Development and Innovation."

47 AGRA, *Strategy Overview for 2017–2021.*

48 AGRA, *African Agriculture Status Report.*

49 AGRA, "Program Development and Innovation."
50 Michel Petit, *The Benefits of Modern Agriculture.*
51 Fukuda-Parr, *The Gene Revolution.*
52 ISAAA, "ISAAA AfriCenter"; ISAAA "ISAAA in 2016."
53 Schurman, "Building an Alliance for Biotechnology in Africa."
54 Schnurr, "Biotechnology and Bio-Hegemony in Uganda," 641; Newell, "Bio-Hegemony."
55 ABSP II, "What Is ABSPII?"
56 PBS, "About Us"; USAID, "$3 Billion in Private Sector Investment."
57 ABNE, "ABNE Brochure."
58 Matunhu, "A Critique of Modernization and Dependency Theories in Africa"; Haakonsson, Gammelgaard, and Just, "Corporate Scramble for Africa," 5.
59 Moyo, Yeros and Jha, "Imperialism and Primitive Accumulation."
60 Holt-Giménez, "Out of AGRA."
61 Patel, "The Long Green Revolution," 40.
62 McKeon "The New Alliance for Food Security and Nutrition," 4; McMichael, "The Land Grab and Corporate Food Regime Restructuring," 683.
63 Bishop and Green, "Philanthro-Capitalism"; Sulle and Hall, "Reframing the New Alliance Agenda."
64 Interview with Informant 33, Kampala, 15 June 2010. See also Dano (2007) and Schurman (2017).
65 Holt-Giménez, "Out of AGRA."
66 Judith Oduol, email message to author, 13 May 2015.
67 Masiga, email message to author, 23 June 2013.
68 *Larouse Agricole*, in Andersson and Sumberg, "Knowledge Politics in Development-Oriented Agronomy," 3.
69 Balch, "Are We Able to Have a Rational Debate about GM?"; Gerson, "Are You Anti-GMO?"; Lynas, *Seeds of Science.*
70 Andersson and Sumberg, "Knowledge Politics in Development-Oriented Agronomy," 5.
71 Scoones and Thompson, "The Politics of Seed in Africa's Green Revolution," 4.
72 Yapa, "Improved Seeds and Constructed Scarcity"; Kloppenburg, *First the Seed.*
73 Pehu and Ragasa, *Agricultural Biotechnology*; Qaim, "Benefits of Genetically Modified Crops"; Just and Kaiser, "A Good Deal for the Environment and the Poor."
74 Dowd-Uribe, "Engineering Yields and Inequality," 162.

75 Gray and Dowd-Uribe, "A Political Ecology of Socio-Economic Differentiation."
76 Bernstein, "Considering Africa's Agrarian Questions"; Andersson and Sumberg, "Knowledge Politics in Development."
77 Zimmerer and Bassett "Approaching Political Ecology."
78 Carney, "Converting the Wetlands, Engendering the Environment"; Schroeder, "Shady Practice"; Flachs, "Redefining Success"; Galt, "Placing Food Systems in First World Political Ecology"; Nyantakyi-Frimpong and Bezner Kerr, "A Political Ecology of High-Input Agriculture"; Cafer and Rikoon, "Coerced Agricultural Modernization"; Jarosz, "Growing Inequality"; Ferguson, *The Anti-Politics Machine*.
79 Perreault, Bridge, and McCarthy, *Routledge Handbook of Political Ecology*, 9.
80 Ibid, 9.
81 Robbins, *Political Ecology: A Critical Introduction*. Critical Introductions to Geography.
82 Fischer et al., "Social Impacts of GM Crops in Agriculture," 8611.

CHAPTER ONE

1 Schnurr, "Breeding for Insect-Resistant Cotton."
2 Rooney, "Sorghum Improvement."
3 Nasser and Ortiz, "Cassava Improvement"; Hahn and Keyser, "Cassava: A Basic Food of Africa."
4 USDA, *Agricultural Statistics 1960*, 33.
5 Byerlee, "Modern Varieties, Productivity, and Sustainability," 698.
6 Kloppenburg, *First the Seed*, 11.
7 Ibid., 11.
8 Zerbe, "Seeds of Hope, Seeds of Despair"; Derera and Musimwa, "Why SR52 Is Such a Great Maize Hybrid?"
9 Kloppenburg "First the Seed," xvi.
10 Xu and Crouch, "Marker-Assisted Selection in Plant Breeding."
11 Serraj et al., "Recent Advances in Marker-assisted Selection"; Lau et al., "Advances to Improve the Eating and Cooking Qualities of Rice"; Bimpong et al., "Improving Salt Tolerance of Lowland Rice Cultivar."
12 Rudi et al., "Economic Impact Analysis of Marker-Assisted Breeding."
13 Ibid.
14 Namanda et al., *Micropropagation and Hardening Sweetpotato Tissue Culture Plantlets*, 3.

15 Okogbenin, Egesi, and Fregene, "Molecular Markers and Tissue Culture," 41.

16 Njuguna et al., "Socio-Economic Impact of Tissue Culture Banana"; Kabunga, Dubois, and Qaim, "Impact of Tissue Culture Banana Technology."

17 Interview with Informant 29, Kampala, 2 June 2015.

18 ISAAA, *Agricultural Biotechnology.*

19 Wu et al., "Unintended Consequence of Plant Transformation"; Rivera et al., "Physical Methods for Genetic Plant Transformation."

20 Rivera et al., "Physical Methods for Genetic Plant Transformation"; Gao and Nielsen, "Comparison between Agrobacterium-Mediated and Direct Gene Transfer Using the Gene Gun."

21 Robert Goodman cited in Kloppenburg, "First the Seed," 311; Qaim, *Genetically Modified Crops and Agricultural Development,* 41.

22 Over the past ten years plant breeders have increasingly moved away from single event GM breeding towards gene pyramiding that allows for the stacking of multiple GM traits within a single transformation event.

23 ISAAA, *Agricultural Biotechnology.*

24 Schnurr, "GMO 2.0: Genetically Modified Crops and the Push for Africa's Green Revolution."

25 Abrahams, Mchiza, and Steyn, "Diet and Mortality Rates in Sub-Saharan Africa," 801.

26 Michael Lipton in Herring, "The Genomics Revolution and Development Studies," 22.

27 Ronald, "Lab to Farm," e1001878.

28 Li et al., "High-Efficiency TALEN-Based Gene Editing Produces Disease-Resistant Rice"; Voytas, Gao and McCouch, "Precision Genome Engineering and Agriculture," e1001877.

29 Bortesi and Fischer, "The CRISPR/Cas9 System for Plant Genome Editing and Beyond."

30 Xiong, Ding, and Li, "Genome-Editing Technologies," 15019; Khatodia et al., "The CRISPR/Cas Genome-Editing Tool," 506.

31 Cyranoski, "CRISPR Tweak"; Waltz, "CRISPR-Edited Crops," 582.

32 While over 3,000 mutant varieties have been registered worldwide, this breeding technique has only limited penetration in Africa. See Kharkwal and Shu (2010) and Jain (2007).

33 Qaim, *Genetically Modified Crops and Agricultural Development,* 32.

34 Herring, "The Genomics Revolution and Development Studies," 24.

35 Kloppenburg, *First the Seed,* 10.

CHAPTER TWO

1 National Academies of Sciences, Engineering, and Medicine, *Genetically Engineered Crops: Experiences and Prospects*, 12.
2 Zaid et al., *Glossary of Biotechnology for Food and Agriculture.*
3 Mugwagwa and Kiplagat, "Legislative and Policy Issues Associated with GE Crops in Africa," 48.
4 Thomson, "South Africa: An Early Adopter of GM Crops."
5 Ayele, "The Legitimation of GMO Governance in Africa," 244.
6 Nang'Ayo, Simiyu-Wafukho and Oikeh, "Regulatory Challenges for GM Crops in Developing Economies," 2.
7 Rijssen et al., "A Critical Scientific Review."
8 Ayele, "The Legitimation of GMO"; Jaffe, "Ensuring Biosafety at the National Level."
9 Feris, "Risk Management and Liability for Environmental Harm Caused by GMOs."
10 Ayele and Wield, "Science and Technology Capacity Building and Partnership in African Agriculture," 639.
11 Le Chalony and Moisseron, "Research Governance in Egypt," 375.
12 Karembu, Nguthi and Abdel-Hamid, *Biotech Crops in Africa*, 22. See also Abdallah, "GM Crops in Africa."
13 Le Chalony and Moisseron, "Research Governance in Egypt," 385; Ayele and Wield, "Science and Technology Capacity" 641.
14 Le Chalony and Moisseron, "Research governance in Egypt," 379.
15 Madkour, "Egypt: Biotechnology from Laboratory to the Marketplace," 98.
16 The initial research and adaptation of the Bt pesticide was funded through a USAID partnership between AGERI and the University of Wyoming in the US, facilitated by ABSP and USAID-Cairo. Pioneer Hi-Bred then helped to facilitate a domestic patent on the new product, which was distributed through a new domestic private venture. But this pesticide was the lone product to emerge through this co-development channel, which, according to Ayele and Wield (2005, 642), was scuttled by low levels of investment and a weak domestic private sector.
17 Madkour, El Nawawy, and Traynor, *Analysis of a National Biosafety System*, 16.
18 Ibid., 2.
19 Of these two the GM potato progressed the furthest. This transgenic potato was designed by scientists at Michigan State University to resist

the potato tuber moth. Fourteen transgenic lines were planted in Egypt in 1997 under AGERI's care. Five years of field testing were completed, with positive results – resistance was calculated to be between 99 and 100 per cent. But there was no subsequent effort to commercialize these potatoes, and enthusiasm dissipated after the completion of the trials. ABSP surmises that Egyptian officials were concerned about market access: most of Egypt's potatoes are exported to Europe, and, with their ban on GM imports in place, they would have to effectively segregate the Bt potatoes to ensure there was no contamination and risk of losing this valuable market (Brenner 2004, 41).

20 Cohen and Paarlberg, "Unlocking Crop Biotechnology in Developing Countries," 1570.

21 Madkour, El Nawawy, and Traynor, *Analysis of a National Biosafety System*, 20.

22 Ibid., 21.

23 Adenle, Morris and Parayil, "Status of Development, Regulation and Adoption," 162.

24 Carline Brenner, *Telling Transgenic Technology Tales*, 50.

25 Wambugu, "The Importance of Political Will," 9.

26 Cohen and Paarlberg, "Unlocking Crop Biotechnology in Developing Countries," 1570.

27 Interview with Informant 10, Kampala, 6 May 2011.

28 Interview with Informant 39, Entebbe, 9 May 2011.

29 PBS, "About US." Much of this material on this donor-driven program first appeared in Schnurr, "Biotechnology and Bio-Hegemony in Uganda."

30 ABNE, "About Us."

31 ABNE, *ABNE in Africa*. This introduction to ABNE first appeared in Schnurr and Gore (2015).

32 Interview with Informant 10, Kampala, 6 May 2011.

33 Ibid.

34 ABSP II, "Agricultural Biotechnology Support Project II."

35 Interview with Informant 51, Kampala, 9 May 2011.

36 Interview with Informant 12, Kampala, 7 May 2011.

37 Interview with Informant 55, Kampala, 8 May 2012, and interview with Informant 57, 11 May 2012. For more on PBS's "seeing is believing" tours see Schnurr (2013, 649).

38 Interview with Informant 10, Kampala, 6 May 2011.

39 This is an abbreviated version of an argument that appeared in Schnurr and Gore (2015).

40 Interview with Informant 29, Kampala, 27 May 2016. There are many reasons advanced for why the legislative process has continued to stall, including the lack of a free-standing ministry of science and technology, the lack of expertise on the part of the minister responsible, and infighting between the ministries implicated in the bill's passage.

41 Behind the scenes, proponents complain that this latest setback was due in large part to activists who gained an ally in First Lady Janet Museveni, who used her considerable influence to consolidate opposition to the bill (Afedraru and Sekitoleko 2018).

42 Agaba, "Ugandan Scientists Skeptical"; Observer, "Parliament Passes GMO Bill."

43 Lemgo and Timpo, "Case Study: Ghana."

44 Yates, "Ghana: Request for Funds for Biotechnology."

45 Interview with Informant 99, Accra, 20 July 2015.

46 ABNE, *ABNE in Africa*, 11.

47 Ibid., 11.

48 Interview with Informant 99, Accra, 20 July 2015.

49 ABNE, "Supporting Malawi in Overall Regulatory Handling."

50 Lemgo and Timpo, "Case Study: Ghana."

51 Interview with Informant 94, Accra, 6 July 2015. Food Sovereignty Ghana argued that there was a legal precedent for this process of invalidating regulatory approvals based on improper procedure, citing a well-known case where the nation's electoral commission was found by the Supreme Court to be in violation of its mandate, resulting in the suspension of its operations and the invalidating of several contracts that had been negotiated using an incorrect process.

52 Agyeman, "Court Dismisses Injunction on Commercialising of GMOs."

53 Zerbe, "Contesting Privatization."

54 Nelson, "Case Studies of African Agricultural Biotechnology Regulation."

55 Zerbe, "Biodiversity, Ownership, and Indigenous Knowledge."

56 Interview with Informant 39, Entebbe, 9 May 2011.

57 Ibid.

58 Morris, "Biosafety Regulatory Systems in Africa," 68. Opposition to the Model Law was so strong that it was officially reviewed by the African Union in 2007. Some of the more contentious elements of the law were softened, though advocates of genetic modification continued to dismiss this revised version as excessively onerous, especially in its restrictive interpretation of the precautionary principle and its mandate to regulate products derived from GM crops.

59 Southern African Regional Biosafety Network, "Media Has Key Role in African Biosafety."

60 Interview with Informant 99, Accra, 20 July 2015.

61 Interview with Informant 79, Nairobi, 1 June 2016.

62 Clapp, "The Political Economy of Food Aid."

63 Bowman, "Sovereignty, Risk and Biotechnology:"

64 Zerbe, "Feeding the Famine?"

65 As quoted in Andrew Bowman, "Sovereignty, Risk and Biotechnology," 1,374.

66 Ibid., 1369–91.

67 Interestingly, Zambia's decision to ban GM crops has them uniquely positioned to meet the rising demand for non-GM maize. According to the *Epoch Times*, a representative from the World Food Program said that "the fact that Zambia is producing white, non-GM maize has in a way created a niche market because few countries are able to meet the global demand" (Stieber 2013).

68 Stieber, "GMOs, A Global Debate."

69 Shanghai Daily, "Zambia Allows South African Chain Store."

70 Keeley and Scones, *Contexts for Regulations: GMOs in Zimbabwe.*

71 News Day, "Zim Gets $8,1m Food Aid Emergency Fund."

72 News Day, "Africa Takes Fresh Look at GMO Crops."

73 New Zimbabwe, "Zimbabwe: Govt Says No to GMO Imports."

74 Mudukuti, "We May Starve, but at Least We'll Be GMO-Free"; Mudukuti, "Zimbabwe's Rejection of GMO Food Aid Is a Humanitarian Outrage."

75 Gogo, "Zimbabwe: Food Shortages Re-Ignite GMO Concerns."

76 Chambers et al., *GM Agricultural Technologies for Africa.*

77 African Union. *African Model Law on Safety in Biotechnology*, 2001.

78 The United Republic of Tanzania, *Environmental Management (Biosafety) Regulations 2009*, Article 6.

79 Schmickle, "Tanzania Becomes a Battleground in Fight."

80 Wedding and Tuttle, *Pathways to Productivity.*

81 Chambers et al., *GM Agricultural Technologies for Africa*, 22.

82 Mpinga, "State To 'Soften' Bio-Tech Rules"; OFAB Tanzania, "Tanzania: GMOs Good for Agriculture Development, Says Don."

83 Pesa Times, "GMOs in Tanzania, Scientist Wants Review."

84 Tanzania Daily News, "Tanzania: Growing of Biotech Crops Increases, Forum Told."

85 ISAAA, "Tanzania Deputy Minister for Agri."

86 The East African, "Tanzania Plans April Trials for GMO Maize Varieties." This change was lobbied for hard by the BMGF-funded

organizations such as AATF, which was unwilling to proceed with plans for its Water Efficient Maize for Africa (WEMA) program as long as this strict liability penalty remained in place. Soon after the government made this decision AATF announced that it was now ready to move forward with the CFT stage, initiated in April 2016 in Dodoma.

87 Mirondo, "Government Bans GMO Trials."

88 Paarlberg, *Toward a Regional Policy on GMO Crops*; Gachenga, "Status of Biotechnology and Biosafety Policies & Laws in COMESA Countries."

89 These points about super-national regulatory harmonization measures were initially made in Schnurr (2013, 646–48).

90 Interview with Informant 29, Kampala, 19 June 2013. See also Ranneberger, "Cautious Kenya Finally Enacts Long Awaited Biosafety Act."

91 Interview with Informant 80, Nairobi, 12 May 2015.

92 The Séralini study tested the toxicity and health impacts of GM maize NK603 on rats. Rats that had been fed maize NK603 developed tumours, which were linked to the consumption of GM maize. The findings caused a media storm when some of the scientific community and biotech industry heavily scrutinized and discredited Séralini's methodology. They alleged that Séralini used rats that were prone to developing tumours, a decision that Séralini defended. Due to these accusations, the study was retracted from the journal *Food and Chemical Toxicology* and republished the following year in *Environmental Sciences Europe* after minor revisions. For more on this controversy see Casassus (2014), Fagan et al. (2015), and Resnik (2015).

93 Health Minister Beth Mugo quoted in Mwaniki, "GMOs Banned as Cancer Fears Grow." The ban was initially set up to allow Kenyan researchers time to complete a comprehensive study on any potential negative impacts of GM crops. This report was completed in early 2013 but never made public – most surmise that the report confirmed the cabinet's fears, and the ban persisted (interview with Informant 80, Nairobi, 2 June 2016).

94 Interview with Informant 80, Nairobi, 11 May 2015.

95 Ochieng and Matete, "Kenya: Governors Urge State." See also Rawlings, "Lift Genetically Modified Organisms Ban"; USSEC, "Kenyan Governors Call for the Lifting of GMO Ban"; Daily Nation, "Governors: Lift Ban on Genetic Foods."

96 Interview with Informant 80, Nairobi, 2 June 2016.

97 Gebre, "Kenya Says No to Genetically Modified Maize Trials."

98 El Wakeel, "Agricultural Biotechnology and Biosafety Regulations."

99 Ibid.

100 ISAAA, "Sudanese Farmers Reap Benefits from Bt Botton."

101 There are rumblings that Sudan is unhappy with the deal offered by the Chinese, who offered them Bt events within Chinese germplasm (this contrasts with both South Africa and Burkina Faso, where Bt events were backcrossed into local cotton varieties). One continental expert who went in person to view the Sudanese cotton fields in early 2015 reported that Sudanese officials were hoping to renegotiate the deal to allow this backcrossing, in the hopes of retaining a greater share of the profits (interview with Informant 99, Accra, 20 July 2015).

102 Forsyth, *Critical Political Ecology*, 10.

103 Juma and Serageldin, *Freedom to Innovate*, 95–6.

104 Interview with Informant 33, Kampala, 15 June 2010.

105 Mugisha and Afadhali, "Rwanda Wary of GMOs Use in East Africa."

106 Lyatuu, "Uganda: Pressure to Pass GMO Bill Gains Momentum."

CHAPTER THREE

1 Gouse, Kirsten, and Van der Walt, *Bt Cotton and Bt Maize*.

2 Gouse, Kirsten, and Jenkins, "Bt Cotton in South Africa."

3 Jozini Municipality, *Jozini Local Municipality IDP Review*.

4 Schnurr, "Lowveld Cotton."

5 Bennett, Morse, and Ismael, "The Economic Impact of Genetically Modified Cotton on South African Smallholders."

6 Ismael, Bennett, and Morse, "Benefits from Bt Cotton Use by Smallholder Farmers in South Africa."

7 Horsch, "Plant Biotechnology Research in Africa."

8 De Grassi, *Genetically Modified Crops and Sustainable Poverty Alleviation in Sub-Saharan Africa.*

9 Interview with T. J. Buthelezi, Ndumo, 31 January 2005.

10 Shankar, Bennett, and Morse, "Output Risk Aspects," 290.

11 Fok et al., "Contextual Appraisal of GM Cotton Diffusion in South Africa," 480.

12 Cotton South Africa, "Textile Statistics."

13 Morse and Bennett, "Impact of Bt Cotton on Farmer Livelihoods," 227.

14 Fok et al., "Contextual Appraisal of GM Cotton Diffusion," 478.

15 Tripp, *Biotechnology and Agricultural Development Transgenic*, 79.

16 Witt, Patel, and Schnurr, "Can the Poor Help GM Crops?"

17 Ntuli, "Makhathini Cotton Comeback."
18 Ngobese, "Cotton Project Set to Transform Rural Community."
19 It is important to note that largeholder farmers have not followed the same volatile trajectory. Adoption rates continue to hover near 100 per cent, with 95 per cent of farmers now growing the stacked variety that combines insect resistance and herbicide tolerance (ISAAA 2014). Longer-term studies suggest that poorer farmers tended to abandon Bt cotton faster than their wealthier counterparts: most of those who continue to grow Bt cotton are wealthier farmers with greater access to off-farm incomes (Morse and Mannion 2009).
20 Gouse et al., "A GM Subsistence Crop in Africa"; Gouse, Shankar, and Thirtle, "The Decline of Bt Cotton in KwaZulu-Natal"; Hofs et al., "Diffusion du Cotton Génétiquement Modifie."
21 Gouse, Shankar, and Thirtle, "The Decline of Bt Cotton in KwaZulu-Natal," 105.
22 Gouse et al., "A GM Subsistence Crop in Africa."
23 Glover, "Exploring the Resilience of Bt Cotton," 974.
24 Harsh and Smith, "Technology, Governance and Place."
25 Mosse, *Cultivating Development*.
26 Vitale and Greenplate, "The Role of Biotechnology in Sustainable Agriculture," 247.
27 Traore, Hema, and Traore, "Bt Cotton in Burkina Faso," 26.
28 Keetch et al., *Biosafety in Africa*, 19.
29 Traore, Hema, and Traore, "Bt Cotton in Burkina Faso," 31.
30 Roberts, *Two Worlds of Cotton*; Bassett, *The Peasant Cotton*.
31 Vitale, Ouattarra, and Vognan, "Enhancing Sustainability of Cotton Production."
32 Vitale and Greenplate, "The Role of Biotechnology in Sustainable Agriculture," 246.
33 Martin et al., "Pyrethroid Resistance Mechanisms."
34 Traoré et al., *The Effect of Bt Gene on Cotton Productivity*.
35 Interview with Informant 89, Bobo, 1 July 2015.
36 In the colonial period, cotton production in Burkina Faso was facilitated by the French Company for Textile Development, which controlled all aspects of the crop's production from start to finish. This conglomerate persisted for fifteen years following independence, but in 1974 it was transformed into SOFITEX, a parastatal company composed of textile enterprises but controlled largely by state interests. In the 1990s the industry saw the entry of two other companies – one operating in the east and the other in the centre of the country – but SOFITEX remains the largest and most powerful by far, covering over

80 per cent of the country's cotton-growing area (interview with Informant 91, Ouagadougo, 2 July 2015).

37 Interview with Informant 125, phone interview, 21 July 2015.

38 Vitale, Ouattarra, and Vognan, "Enhancing Sustainability of Cotton Production," 1,147

39 Traoré et al., "Testing the Efficacy and Economic Potential of Bollgard II."

40 Interview with Informant 125, phone interview, 21 July 2015.

41 Traoré, Hema, and Traoré, "Bt Cotton in Burkina Faso," 20.

42 Traoré et al., "Testing the Efficacy"; Vitale et al., "Second-Generation Bt Cotton."

43 Vitale et al., "The Commercial Application of GMO Crops in Africa."

44 Vitale and Greenplate, "The Role of Biotechnology in Sustainable Agriculture," 275.

45 Ibid., 279.

46 Ibid., 266.

47 Ibid., 266. This research further showed that pesticide applications were reduced by over two-thirds. See ibid., 282.

48 Interview with Informant 91, Ouagadougo, 2 July 2015.

49 Estur, Quality and Marketing of Cotton Lint in Africa.

50 It is worth noting that late nineteenth and early twentieth century colonial governments originally sought to improve cotton quality characteristics in Africa by introducing American cultivars of cotton. Now, over a century later, American transgenic cultivars are the reason for a decrease in the quality of African cotton.

51 Bassett, The Peasant Cotton; Bingen, "Cotton in West Africa"; Schwartz, "L'évolution de l'agriculture en zone cotonnière."

52 Estur, Quality and Marketing of Cotton Lint in Africa.

53 Interview with Informant 91, Ouagadougo, 2 July 2015.

54 Interview with Informant 125, phone interview, 21 July 2015.

55 Dowd-Uribe, "Engineered Outcomes: The State and Agricultural"; Interview with Informant 91, Ouagadougo, 2 July 2015.

56 The Burkinabè cultivars in use were known to exhibit variance in ginning ratios due to fluctuating growing conditions (Dessauw and Hau, "Cotton Breeding").

57 Interview with Informant 90, Bobo, 1 July 2015; Interview with Informant 91, Ouagadougo, 2 July 2015.

58 EcoBank, "Burkina Faso: Changes Loom for 2015/16 Cotton Season," 2.

59 Interview with Informant 91, Ouagadougo, 2 July 2015.

60 EcoBank, "Burkina Faso: Changes Loom for 2015/16 Cotton Season," 3.

61 Interview with Informant 91, Ouagadougo, 2 July 2015.

62 Interview with Informant 90, Bobo, 1 July 2015.

63 Interview with Informant 90, Bobo, 1 July 2015.

64 Interview with Informant 91, Ouagadougo, 2 July 2015.

65 Bavier, "Burkina Faso Settles Dispute with Monsanto Over GM Cotton."

66 Dowd-Uribe, "Engineering Yields and Inequality?" 163.

67 Ibid., 163

68 Interview with Informant 125, phone interview, 21 July 2015; Interview with Informant 126, phone interview, 27 August 2015.

69 Vitale and Greenplate, "The Role of Biotechnology in Sustainable Agriculture," 255.

70 Interview with Informant 88, phone interview, 11 September 2015.

71 Interview with Informant 91, Ouagadougo, 2 July 2015

72 Bernal, "Cotton and Colonial Order in Sudan."

73 Abdeldaffie, Elhag and Bashir, "Resistance in the Cotton Whitefly"; Eveleens, "Cotton-Insect Control in the Sudan."

74 El-Tigani, El-Amin and Ahmed, "Strategies for Integrated Pest Control in the Sudan."

75 Suliman et al., "Evaluation of Two Chinese Bt Cotton Genotypes," 106. The Sudanese opted not to deal directly with Monsanto, choosing instead to import Chinese varieties. They gambled that, while these foreign varieties would produce lower-quality cotton, the increase in quantity would still make the venture worthwhile. See Bavier, "How Monsanto's GM Cotton Sowed Trouble in Africa" (Bavier 2017).

76 Keetch et al., Biosafety in Africa.

77 Professor Hassan Kanan, agriculture ministry expert in cotton, quoted in personal communication. Joe Bavier, email message to author, 29 November 2017.

78 El-Tigani, El-Amin, and Ahmed, "Strategies for Integrated Pest Control in the Sudan."

79 Hillocks, "Is There a Role for Bt Cotton in IPM for Smallholders in Africa?" 139.

80 Professor Hassan Kanan, agriculture ministry expert in cotton, quoted in personal communication. Joe Bavier, email message to author, 29 November 2017.

81 Ibid.

82 El-Tigani, El-Amin, and Ahmed, "Strategies for Integrated Pest Control in the Sudan," 551.

83 Anderson, Valenzuela, and Jackson, "Recent and Prospective Adoption of Genetically Modified Cotton."

84 The first is Bouët and Gruère, "Refining Opportunity Cost Estimates." The second is Elbehri and Macdonald, "Estimating the Impact of Transgenic Bt Cotton," 2062.

85 Cabanilla, Abdoulaye, and Sanders, "Economic Cost of Non-Adoption of Bt Cotton."

86 Mulwa et al., "Estimating the Potential Economic Benefits," 21.

87 Abdallah, "GM Crops in Africa: Challenges in Egypt."

88 USDA Foreign Agriculture Service, *Egypt: Agricultural Biotechnology Annual 2014*, 5.

89 Interview with Informant 98, 23 July 2015; interview with Informant 125, phone interview, 21 July 2015; interview with Informant 126, phone interview, 27 August 2015. For more on the connections between the trajectory of genetic modification and political changes in Egypt, see chapter 2.

90 Interview with Informant 84, Nairobi, 11 May 2015.

91 Waturu, "The Status of Bt-Cotton Confined Field Trials."

92 Business Daily, "Uhuru Banks on Biotech Cotton for 50,000 Jobs."

93 Ibid.

94 Otieno, "Lift Genetically Modified Organisms." See also Odhiambo, "Govenors: Lift Ban on Genetic Foods."

95 Poulton, "The Cotton Sector in Northern Ghana," 71.

96 United States Department of Agriculture, "Ghana Cotton Production by Year."

97 Poulton et al., "Competition and Coordination in Liberalized African," 525.

98 Interview with Informant 99, Accra, 5 July 2015.

99 Ibid. The Bt cottonseed came from Burkina Faso. Ghana has long lagged behind its neighbours in its capacity for varietal improvement, so it is no surprise that officials decided to import seed from the Burkina Faso base system that is much more entrenched and sophisticated. See Poulton et al., "Competition and Coordination in Liberalized African," 531.

100 Okine, "Trial of New Cotton Variety."

101 Minister of Agriculture Fiifi Fiavi Kwetey, quoted in Ghana News Agency, "Cotton Development Authority Board."

102 Interview with Informant 99, Accra, 5 July 2015.

103 Early efforts of this industry-wide restructuring have sputtered. Cotton input increased in 2011/12 from 2,500 hectares to 21,000 hectares under cultivation, to 28,000 hectares in 2012/13, but

the collapse of two of the major cotton buyers caused production figures to return to 2,500 hectares in 2014/15 (Ministry of Food and
Agriculture 2015).

104 Bouët and Gruère, "Refining Opportunity Cost Estimates," 262.
105 Schnurr, "Breeding for Insect-Resistant Cotton," 229.
106 Baffes, "The 'Full Potential' of Uganda's Cotton Industry."
107 Hillocks, "Is There a Role for Bt Cotton in IPM," 132; Baffes, "The
     'Full Potential' of Uganda's Cotton Industry," 67.
108 Ibid., 70.
109 Interview with Informant 2, Kampala, 2 June 2009.
110 Baffes, "The 'Full Potential' of Uganda's Cotton Industry," 71.
111 Interview with Informant 2, Kampala, 5 June 2010.
112 Baffes, "The 'Full Potential' of Uganda's Cotton Industry," 67.
113 Horna, Zambrano, and Falck-Zepeda, *Socioeconomic Considerations
     in Biosafety Decisionmaking*, 25.
114 Jo Baffes, "The 'Full Potential' of Uganda's Cotton Industry," 75.
115 Interview with Informant 60, Kampala, 15 May 2012.
116 Interview with Informant 47, Kampala, 4 May 2011.
117 Schnurr and Gore, "Getting to 'Yes,'" 59.
118 Horna, Zambrano, and Falck-Zepeda, *Socioeconomic Considerations
     in Biosafety Decisionmaking*, 38
119 Interview with Informant 60, Kampala, 14 June 2013.
120 Wamboga-Mugirya, "Biotech Can Improve Agric Productivity," 5.
121 Interview with Informant 12, Kampala, 6 June 2009.
122 Interview with Informant 66, Kampala, 11 May 2011.
123 Interview with Informant 60, Kampala, 15 May 2012.
124 Ssonko, "Monsanto Chief Points at Roadblocks."
125 Baffes, "The 'Full Potential' of Uganda's Cotton Industry," 81.
126 Horna, Zambrano, and Falck-Zepeda. *Socioeconomic Considerations
     in Biosafety Decisionmaking*, 145–6.
127 Ibid., 70.
128 Interview with Informant 60, Kampala, 14 June 2013.
129 Horna, Zambrano, and Falck-Zepeda, *Socioeconomic Considerations
     in Biosafety Decisionmaking*, 75.
130 Interview with Informant 47, Kampala, 4 May 2011.
131 Baffes, "The Cotton Sector of Uganda," 16.
132 Horna, Zambrano, and Falck-Zepeda, *Socioeconomic Considerations
     in Biosafety Decisionmaking*, 79.
133 Interview with Informant 60, Kampala, 15 May 2012. Cotton
     experts advised against this practice of intercropping, arguing that

this practice exacerbates pest losses by allowing sucking pests such as aphids and lygus to hide in the canopy, but it remains the favoured cropping strategy among the vast majority of farmers we consulted.

134 Interview with Informant 60, Kampala, 14 June 2013.

135 Horna, Zambrano, and Falck-Zepeda. *Socioeconomic Considerations in Biosafety Decisionmaking*, 81.

136 Ibid., 92.

137 Idbid., 147. The actual figure of the technology fee will only be fixed once GM cotton is ready for commercial release in Uganda. Regardless of its specific amount, the negative impacts of a technology fee are likely to be more pronounced in Uganda than it was in either South Africa or Burkina Faso. Uganda cotton growers have enjoyed a long history of seed subsidization by the CDO: until 2011, up to five kilograms of cottonseed was distributed for free by the CDO, and seed is commonly perceived as a "free good" (Horna, Zambrano, and Falck-Zepeda 2013, 57). This practice was phased out in 2011, at which point the CDO began charging USh 3,000 for three kilograms of seed (enough to seed one acre of cotton). While farmers maligned this sudden increase, industry officials note that seed remains heavily subsidized and is "very cheap" for farmers (interview with Informant 47, Kampala, 4 May 2011).

138 Baffes, "The 'Full Potential' of Uganda's Cotton Industry," 73.

139 Interview with Informant 60, Kamapala, 14 June 2013.

140 Horna, Kyotalimye, and Falck-Zepeda. "Cotton Production in Uganda."

141 Horna, Zambrano and Falck-Zepeda, *Socioeconomic Considerations in Biosafety Decisionmaking*, 29.

142 Smale et al., *Measuring the Economic Impacts of Transgenic Crops*, 88.

143 See Hillcocks et al. (2005, 139), who argue convincingly that Bt cotton works well in agro-ecological systems where Lepidopteran pests dominate, such as China and Mexico, but is not nearly as well suited in African environments, where a mix of pests is the norm.

144 Glover, "Is Bt Cotton a Pro-Poor Technology," 484.

CHAPTER FOUR

1 FAO, "Statistics Division."

2 BMGF, "Our Big Bet for the Future."

3 BMGF, "Awarded Grands."

4 McCann, *Maize and Grace*, 210.
5 Byerlee and Heisey, "Past and Potential Impacts of Maize Research."
   See also Smale and Jayne (2003) for a thorough history of maize
   breeding in eastern and southern Africa.
6 Byerlee and Heisey, "Past and Potential Impacts of Maize Research,"
   261.
7 McCann, *Maize and Grace*, 149.
8 Ibid., 154.
9 Other dangers associated with hybrid breeding include the creation
   of a genetically homogenous population that is less resilient than a
   genetically diverse population (Jacobson, *From Betterment to Bt*, 35)
   and an erosion of both local knowledge and diversity of longstand-
   ing, locally adapted landraces (McCann, "Political Ecology of Cereal
   Seed," 31).
10 McCann, *Maize and Grace*, 147.
11 Jacobson, *From Betterment to Bt*, 34. For instance, Kenya's late colonial policy emphasized research into maize well suited to African
   farming systems, while similar prioritizations were incorporated into
   experimental programs in Zambia in the post-independence period
   and in Zimbabwe after 1980. See also McCann, *Maize and Grace*,
   172).
12 Gouse et al., "A GM Subsistence Crop in Africa."
13 Ibid., 84. In addition to these direct losses, stem borer damage also has
   significant indirect effects: the insects cause damage to the cobs, leading to conditions that allow for the proliferation of mycotoxins (fungal
   toxins that are difficult to detect and present a major danger to human
   health).
14 Keetch et al., "Bt Maize for Small Scale Farmers: A Case Study," 1507.
15 Gouse et al., "A GM Subsistence Crop in Africa."
16 Gouse, Kirsten, and Van der Walt, *An Evaluation of Direct and
   Indirect Impact*, 83.
17 Ibid.
18 Another important factor here was limited seed availability due to
   problems with seed multiplication. Gouse ("GM Maize as Subsistence
   Crop," 164) suggests that the low adoption rates that characterized
   these first few growing seasons were a factor of low supply rather than
   low demand.
19 For a detailed account of this rollout see Gouse et al., "Three Seasons
   of Subsistence Insect-Resistant Maize," 16–19.
20 Gouse et al., "Assessing the Performance of GM Maize," 79.

21 See Gouse et al., "Assessing the Performance of GM Maize," for the original production figures, and Regier et al., "Impact of Genetically Modified, Maize, 370, for the explanation regarding the labour-saving properties of HT maize.

22 Gouse, *Seed Technology and Production System Comparisons*, 7.

23 Ibid., 10.

24 Keetch, *Biosafety in Africa: Experiences and best practices*, 17.

25 Gouse et al., "Genetically Modified Maize."

26 Gouse et al., *Bt Cotton and Bt Maize*; Keetch et al., *Biosafety in Africa*, 17.

27 Gouse, *Seed Technology and Production System Comparisons*, 2; Keetch et al., *Biosafety in Africa*, 17.

28 Gouse, Kirsten, and Van der Walt, *Bt Cotton and Bt Maize*, 11.

29 Department of Agriculture, Forest and Fisheries, "Maize Profile."

30 Gouse, Kirsten, and van der Walt, *Bt Cotton and Bt Maize*.

31 Even the primary author of the most optimistic assessments of South Africa's experiences with GM maize acknowledges that "variability between seasons" is one of the most important trends over a long-term study of farmer experiences with this technology (Gouse, "GM Maize as Subsistence Crop," 164).

32 Van Rensburg et al., "The Influence of Rainfall on the Seasonal Abundance."

33 Gouse, Kirsten, and Van der Walt, *Bt Cotton and Bt Maize*.

34 For more on the endemic levels of high interannual variability in South Africa, see Tyson and Dyer, "Mean Annual Fluctuations"; Preston-Whyte and Tyson, *The Atmosphere and Weather of Southern Africa*, 260; and Mason, "El Niño, Climate Change, and the Southern African Climate."

35 Gouse et al., "A GM Subsistence Crop in Africa," 86.

36 Jacobson and Myhr, "GM Crops and Smallholders," 117.

37 Byerlee and Heisey, "Past and Potential Impacts of Maize Research"; Waddington et al., "Extent and Causes of Low Yield in Maize." This high variability in precipitation had implications for HT adopters as well. The most significant gains were reported in the 2006/07 season, when yields from HT maize in northern KwaZulu-Natal were 184 per cent higher than those from conventional hybrids. The pattern of precipitation played an important role here: Gouse ("GM Maize as Subsistence Crop," 172) suggests that very heavy rains early in the growing season and comparatively little rain later on exacerbated the impact weeds had on yields, impeding yields in non-HT-tolerant

varieties. Thus, there is anecdotal evidence that the high levels of variability in precipitation will similarly exacerbate annual variation among HT varieties.

38 The price differentials for BR maize are even starker, as the technology fee is double that levied for Bt and HT varieties. For specific data on prices see Gouse, *Seed Technology*, and Fischer et al., "Is Bt Maize Effective."

39 Gouse et al., "Three Seasons of Subsistence Insect-Resistant Maize in South Africa?"

40 Soleri et al., "Understanding the Potential Impact of Transgenic Crops," 156.

41 Cleveland and Soleri, "Rethinking the Risk Management Process"; Soleri et al., "Understanding the Potential Impact of Transgenic Crops," 156; Soleri et al., "Testing Assumptions Underlying Economic Research."

42 Foti et al., "Farmer Participatory Screening"; Jacobson, "From Betterment to Bt."

43 Gouse, Kirsten, and Van der Walt. *Bt Cotton and Bt Maize*, 70.

44 Assefa and Van den Berg, "Genetically Modified Maize," 218.

45 Gouse, *Seed Technology and Production System Comparisons*, 5.

46 See Jacobson and Mhyr, "GM Crops and Small Holders," for the case in Eastern Cape, and Fischer et al., "Is Bt Maize Effective," for the case in Limpopo.

47 Kruger, Van Rensburg, and Van Den Berg, "Resistance to Bt Maize in Busseola Fusca."

48 Assefa and Van den Berg, "Genetically Modified Maize."

49 Fischer and Hajdu, "Does Raising Maize Yields Lead to Poverty Reduction?" 310.

50 Assefa and Van den Berg, "Genetically Modified Maize."

51 Stone, "Agricultural Deskilling."

52 Kruger, Rensburg, and Van Den Berg, "Perspective on the Development of Stem Borer Resistance."

53 Ibid., 687. Note that a new stewardship program was initiated during the 2008/09 growing season that included growing education programs and third-party inspections, which seem to have increased compliance rates with refuges (Kruger et al., "Resistance to Bt Maize," 48).

54 Kruger, Rensburg, and Van Den Berg, "Perspective on the Development of Stem Borer Resistance."

55 Monsanto, *User Guide for the production of YieldGard*, 12; Viljoen and Chetty, "A Case Study of GM Maize."

56 Moyo, *Land Ownership Patterns and Income Inequality in Southern Africa.*

57 Assefa and Van den Berg, "Genetically Modified Maize."

58 Iversen et al., "Detection of Transgenes in Local Maize," 10; Fischer and Hajdu, "Does Raising Maize Yields Lead to Poverty Reduction?"

59 Ibid., 310.

60 Iversen et al., "Detection of Transgenes in Local Maize," 10.

61 Ibid.

62 Tabashnik, Brévault, and Carrière, "Insect Resistance to Bt Crops."

63 Van Rensburg, "First Report of Field Resistance."

64 Kruger, Van Rensburg, and Van Den Berg, "Perspective on the Development."

65 Tabashnik, Brévault, and Carrière, "Insect Resistance to Bt Crops," 517.

66 Tabashnik, Rensburg and Carrière, "Field-Evolved Insect Resistance to Bt Crops"; Tabashnik et al., "Defining Terms for Proactive Management."

67 Campagne et al., "Dominant Inheritance of Field-Evolved Resistance."

68 Van Den Berg, Hilbeck, and Bøhn, "Pest Resistance to Cry1Ab Bt Maize."

69 Tabashnik, Rensburg, and Carrière, "Field-Evolved Insect Resistance to Bt Crops"; Tabashnik et al., "Defining Terms for Proactive Management."

70 Binimelis, Pengue, and Monterroso, "Transgenic Treadmill."

71 Gouse, "GM Maize as Subsistence Crop," 171.

72 Fischer and Hajdu, "Does Raising Maize Yields Lead to Poverty Reduction?"

73 ACB, *Africa Bullied to Grow Defective Bt Maize,* 19.

74 De Groote, "Crop Biotechnology in Developing Countries."

75 Interview with Informant 111, phone interview, 2 May 2017.

76 Assem, "Opportunities and Challenges of Commercializing Biotech," 39.

77 Adenle, Morris, and Parayila, "Status of Development, Regulation and Adoption," 162. The absence of a formal national regulatory framework was a source of some contention. Some stakeholders report that a biosafety law was pushed by Monsanto, an accusation that Monsanto's representatives firmly deny.

78 Abdallah, "GM Crops in Africa: Challenges in Egypt," 117.

79 But commercialization was limited to maize produced for animal feed (not human consumption), and the project was only allowed to introduce the event into yellow maize.

80 Assem, "Opportunities and Challenges of Commercializing Biotech," 44.
81 Karembu, Nguthi, and Abdel-Hamid, *Biotech Crops in Africa: The Final Frontier.*
82 USDA Foreign Agriculture Service, *Egypt: Agricultural Biotechnology Annual 2014*, 3.
83 The USDA cites a failure on the part of Egyptian scientists to employ various "risk communication strategies and better communicate with policy makers, media and the public" (USDA, Egypt, 4).
84 Sarant, "Egypt's Legal Battle to Regulate Monsanto's GMOs."
85 USDA Foreign Agriculture Service, *Egypt: Agricultural Biotechnology Annual 2014*, 1.
86 Ibid., 2.
87 See for example, Abdo, Barbary, and Shaltout, "Chemical Analysis of BT Corn 'Mon-810'"; Gab-Alla et al., "Morphological and Biochemical Changes in Male Rats"; Abdo, Barbary, and El-Sayed Shaltout, "Feeding Study with Bt Corn"; El-Shamei et al., "Histopathological Changes in Some Organs of Male Rats." There are a number of legitimate critiques of this research, including small sample sizes and the selection of mice strain. This research is also published in questionable peer-reviewed journals, most of which seem to operate on a fee-for-publication basis. Still, the concerns raised in these experiments certainly deserve more systematic and rigorous replication.
88 USDA Foreign Agriculture Service, *Egypt: Agricultural Biotechnology Annual 2014*, 3.
89 Doss, "Environmentalists Protest Against GM Seeds"; Hussein, "Egyptian Activists Launch First Protest"; Hanna, "Video: Egypt Activists Protest Monsanto's GM Seeds."
90 Olsen, "Monsanto in Egypt." You can view the full video, in Arabic, at CBC Egypt, "#Zayelshams - يز - سمشلا - روذبلا ةلدعملا وراثيأ فيها سما قاتل."
91 The Arab Spring also precipitated a complete overhaul within the institutions responsible for facilitating and regulating GM crops. The new minister of agriculture reconstituted the NBC, recruiting new members drawn from a wider array of sectors invested in the debate including ministries of commerce, health, and foreign affairs; various universities and research centres; and the environmental monitoring agency (Abdallah, "Biatech Crops").
92 Quoted in Ezezika and Daar, "Building Trust in Biotechnology Crops in Light of the Arab Spring," 3.
93 Ibid.
94 USDA Foreign Agriculture Service, "Egypt," 19.

95 De Groote et al., "Assessing the Impact of Bt Maize."

96 Schimmelpfennig, Rosen, and Pray, *Genetically Engineered Corn in South Africa.*

97 Bøhn et al., "Co-Existence Challenges," offer a useful sketch of the challenges of integrating GM maize among small-scale producers in Zambia.

98 McCann, "The Political Ecology of Cereal Seed Development," 32.

99 Diallo and Pixley, CIMMYT *in Sub-Saharan Africa.*

100 Fischer, Van Den Berg, and Mutengwa, "Is Bt Maize Effective in Improving," 2.

101 Though DTMA breeders do recommend that farmers replace OPV seed every three years to prevent the dilution of favourable traits (interview with Informant 85, Nairobi, 10 May 2015).

CHAPTER FIVE

1 Spielman and Zambrano, "Policy, Investment, and Partnerships"; Stads and Beintema, "Agricultural R&D Expenditure in Africa."

2 Spielman, Hartwich, and Grebmer, "Public-Private Partnerships and Developing."

3 Horsch and Montgomery, "Why We Partner," 82.

4 Ibid., 82. See also Spielman, Hartwich, and Grebmer, "Public-Private Partnerships and Developing."

5 Ibid., 261–76.

6 Ibid., 262.

7 See calls in Assefa, "The Dire Need to Support 'Orphan Crop' Research," 8–9; Tadele and Assefa, "Increasing Food Production in Africa"; Naylor et al., "Biotechnology in the Developing World"; and Varshney et al., "Can Genomics Boost Productivity of Orphan Crops," 1172.

8 Haakonsson et al., "Corporate Scramble for Africa."

9 ACB, AFSA, et al., "Modernising African Agriculture."

10 Oxfam, *Moral Hazard? "Mega" Public-Private Partnerships in African Agriculture.*

11 Wambugu, "Development and Transfer of Genetically Modified."

12 Ibid.; GRAIN, "USAID: Making the World Hungry for GM Crops."

13 Interview with Informant 113, Nairobi, 15 May 2017.

14 Odame, Kameri-Mbote, and Wafula, "Governing Modern Agricultural Biotechnology."

15 Qaim, *The Economic Effects of Genetically Modified Orphan Commodities*, 20.

16 Wambugu, "Development and Transfer of Genetically Modified," 112.
17 Cook, "While the West Debates the Ethics."
18 Odame, Kameri-Mbote and Wafula, "Innovation and Policy Process."
19 deGrassi, *Genetically Modified Crops and Sustainable Poverty Alleviation*, 6.
20 Wambugu, "Development and Transfer of Genetically Modified."
21 Gathura, "GM Technology Fails Local Potatoes."
22 deGrassi, *Genetically Modified Crops and Sustainable Poverty Alleviation*, 15.
23 Ojanji, "Partners Recognize Achievements."
24 Bürgi, *Insect-Resistant Maize*.
25 Mabeya and Ezezika, "Unfulfilled Farmer Expectations," S6.
26 Mugo, et al., "Developing Bt Maize for Resource-Poor Farmers."
27 Bürgi, *Insect-Resistant Maize*, 172.
28 Ibid., 171.
29 Ibid., 227.
30 Ibid., 223–5.
31 Interview with Informant 111, phone interview, 2 May 2017.
32 Mugo et al., "Developing Bt Maize for Resource-Poor Farmers," 1493.
33 It remains unclear why IRMA breeders were under the impression that they were operating with public-licence technology for almost seven years, without realizing that any commercialization of IRMA maize would inevitably require access to Monsanto's proprietary licence. In a 2017 interview the former IRMA project leader explains, "the events that we were testing at IRMA, the public events, we believed that ... farmers could access them easier ... we had no agreement with Ottawa, that it would be publicly available, but we believed that they would be." The biochemist at the University of Ottawa who synthesized the Cry genes in question was adamant that he has no proprietary rights over these constructs: "I never applied for a patent, ever, on Bt, because I've never discovered a Bt sequence and I've never invented any technology. I have nothing to patent ... I was not, as a biotechnology professor, going to try to invent something. I didn't invent anything. There's nothing to invent here." Regardless of its source, this mix-up only came to light in 2002 (interview with Informant 111, phone interview, 2 May 2017).
34 Interview with Informant 112, phone interview, 25 May 2017.
35 Bürgi, *Insect-Resistant Maize*, 220.
36 Ibid., 237.

37 Marianne Banziger, quoted in Bürgi, *Insect-Resistant Maize*, 249.
38 De Groote et al., *Assessing the Impact of Bt Maize in Kenya*, 8.
39 KARI and CIMMYT, *Insect Resistant Maize for Africa*, 90.
40 Bürgi, *Insect-Resistant Maize*, 221.
41 Ibid., 251.
42 Willy De Greef, quoted in Bürgi, *Insect-Resistant Maize*, 236.
43 Bürgi, *Insect-Resistant Maize*, 237.
44 Marianne Banziger, quoted in Bürgi, *Insect-Resistant Maize*, 237.
45 Ojanji, "Partners Recognize Achievements in Insect-Resistant Maize Delivery."
46 Syngenta Foundation, "Insect Resistant Maize for Africa – IRMA."
47 Ojanji, "Partners Recognize Achievements in Insect-Resistant Maize Delivery."
48 Loebenstein, "Control of Sweet Potato Virus Diseases."
49 Wambugu, "Development and Transfer of Genetically Modified," 112.
50 Karyeija et al., "Synergistic Interactions of a Potyvirus."
51 Cipriani et al., "Transgene Expression of Rice Cysteine Proteinase Inhibitors," 270.
52 Naylor et al., "Biotechnology in the Developing World," 28–9.
53 Ibid., 32.
54 From Qaim, *Economic Effects of Genetically Modified Orphan Commodities*; Naylor et al., "Biotechnology in the Developing World," 35.
55 Gasura, Mashingaidze, and Mukasa, "Occurrence, Prevalence and Implications of Sweet Potato Recovery"; Rwegasira, "Prevalence of Sweetpotato Virus Vectors."
56 Mwanga et al., "'NASPOT 11,' a Sweetpotato Cultivar"; Mwanga et al., "'NASPOT 7', 'NASPOT 8', 'NASPOT 9.'"
57 See, for instance, Vetten et al., "Serological Detection and Discrimination"; and Hoyer et al., "Identification of the Coat Protein Gene of a Sweet Potato."
58 Tairo et al., "Unravelling the Genetic Diversity"; Karyeija et al., "Variability of Sweetpotato Feathery Mottle Virus; "Karyeija, Gibson, and Valkonen, "Resistance to Sweet Potato Virus Disease"; Karyeija et al., "Synergistic Interactions of a Potyvirus."
59 New Scientist, "Monsanto Failure," 7.
60 Ojanji, "Partners Recognize Achievements."
61 Willy De Greef, cited in Bürgi, *Insect-Resistant Maize*, 236.
62 Odame, Kameri-Mbote, and Wafula, *Governing Modern Agricultural Biotechnology in Kenya*.

63 World Bank, *Uganda: Promoting Inclusive Growth*, 38.
64 Scoones and Thompson, "Politics of Seed in Africa's Green Revolution," 6.
65 Odame, Kameri-Mbote, and Wafula, "Innovation and Policy Process"; Odame, Kameri-Mbote, and Wafula, *Governing Modern Agricultural Biotechnology in Kenya*.
66 Wambugu, "Development and Transfer of Genetically Modified Virus," 112. See also Odame, Kameri-Mbote, and Wafula, "Innovation and Policy Process," 2775.
67 Ibid., 2775. See also CIMMYT, "Honing Skills in Scientific Writing for Publishing."
68 Interview with Informant 113, Nairobi, 15 May 2017.
69 GRAIN, "USAID: Making the World Hungry for GM Crops," 4.
70 Wambugu, "Development and Transfer of Genetically Modified," 113.
71 Harsh and Smith, "Technology, Governance and Place," 256.
72 Interview with Informant 113, Nairobi, 15 May 2017.
73 Harsh and Smith, "Technology, Governance and Place," 253.
74 Wambugu, "Development and Transfer of Genetically Modified," 113.

CHAPTER SIX

1 FAO, FAOSTAT.
2 Smale, Byerlee, and Jayne, *Maize Revolutions in Sub-Saharan Africa*, 3.
3 Shiferaw et al., "Crops That Feed the World," 309.
4 Smale, Byerlee, and Jayne, *Maize Revolutions in Sub-Saharan Africa*, 4.
5 BMGF, "Our Big Bet for the Future: 2015 Gates Annual Letter."
6 All of these figures come from FAOSTAT. Note that the average yields for both 1960 and 2014 exclude South Africa, because, as recounted in chapter 4, its maize yields are quite exceptional across the continent.
7 All of these figures are from Smale et al., "Maize Revolutions in Sub-Saharan Africa," 4. It is important to stress, as the authors do in their original reporting of these figures, that much of this difference in yields stems from differences in production systems – that is, SSA farmers tend to operate on smaller farms, and have less access to fertilizer and irrigation than do their international comparators.
8 Tumusiime et al., "The Cost of Coexistence."
9 AATF, "About Us."
10 Interview with Informant 8, Nairobi, 22 June 2009.
11 Hoag, "Biotech Firms Join Charities," 246.
12 Conway, "Green Revolution to the Biotechnology Revolution."

13 Schurman, "Building an Alliance for Biotechnology in Africa."
14 Boadi and Bokanga, "African Agricultural Technology Foundation Approach"; Mignouna et al., "Delivery of Agricultural Technology to Resource-Poor Farmers."
15 Brooks et al., *Environmental Change and Maize Innovation in Kenya.*
16 Brooks, "Biofortification."
17 Brooks et al., "Environmental Change and Maize."
18 AATF, "Project Brief."
19 Lumpkin and Armstrong, *Staple Crops, Smallholder Farmers and Multinationals.*
20 This includes the Kenya Agricultural Research Institute (KARI) in Kenya, the National Agricultural Research Organisation (NARO) in Uganda, the Commission for Science and Technology (COSTECH) in Tanzania, the Agricultural Research Council (ARC) in South Africa, and the Agricultural Research Institute of Mozambique (IIAM) in Mozambique. See AATF, "Project Brief."
21 Interview with Informant 86, Nairobi, 16 May 2017.
22 African Centre for Biodiversity, *Profiting from the Climate Crisis.*
23 MON 87460 also includes "an npII expression cassette flanked by two functional loxP sites," which, according to Myhr and Quist (*Impact Assessment*, 3), has the potential to accelerate the pace of bacterial resistance to antibiotics.
24 An important distinction to note here is that between drought tolerance and water use efficiency, which are often used interchangeably and misunderstood as the same trait. Both traits relate to yield potential under conditions of water stress. Water use efficiency is the ratio between yield and crop water use, often understood as the goal of "more crop per drop" (Blum, "Effective use of Water," 119). Drought tolerance is an umbrella term that describes techniques used by crops to withstand dry weather, including both dehydration avoidance and dehydration tolerance.
25 Castiglioni et al., "Bacterial RNA Chaperones Confer Abiotic Stress."
26 Gurian-Sherman, *High and Dry*, 2.
27 USDA, *Monsanto Company Petition*, 33.
28 Abate et al., "Characteristics of Maize Cultivars in Africa," 5.
29 Conrow, "Tanzania Plants Its First GMO Research Crop."
30 Very little material exists that discusses the inclusion of the Bt gene in WEMA. Most sources do not mention it or do not question the inclusion or rationale behind this decision. The only exception is the African Centre for Biodiversity (ACB) (*Africa Bullied to Grow*;

Profiting from the Climate Crisis). The ACB shows how this gene, which was imported from the United States, has already failed to produce positive results in South Africa, leading them to conclude that Africa was bullied into growing "defective Bt Maize" (ACB, *Africa Bullied to Grow*). They suggest that the inclusion of the gene was kept quiet, which is corroborated by the lack of available sources about this event.

31 Kireger, Kenya Agricultural and Livestock Research Organization.

32 Interview with Informant 85, Nairobi, 31 July 2015.

33 The longer-term vision was to incorporate weed resistance alongside these two existing traits (interview with Informant 85, Nairobi, 31 May 2016).

34 In South Africa, the only country with a vibrant large-scale maize sector, only farmers with holdings of under three hectares will be exempt from royalty payments.

35 Whitfield, *Adapting to Climate Uncertainty in African Agriculture*.

36 AATF, "Efforts Underway to Develop Maize."

37 Oikeh et al., "The Water Efficient Maize for Africa Project."

38 Lumpkin and Armstrong, "Staple Crops, Smallholder Farmers and Multinationals."

39 Tefera et al., "Resistance of Bt-Maize," 206.

40 Mulaa et al., "Evaluation of Stem Borer Resistance Management."

41 De Groote et al., "Assessing the Potential Economic Impact."

42 Kostandini, Mills, and Mykerezi, "Ex-Ante Evaluation of Drought Tolerant Varieties."

43 Kostandini, Rovere, and Abdoulaye, "Potential Impacts of Increasing Average."

44 Nagarajan et al., "Political Economy of Genetically Modified Maize," 208.

45 Kent, "Bringing Drought Tolerance to Africa."

46 CIMMYT, "High Expectations Among Stakeholders."

47 Brooks et al., *Environmental Change and Maize Innovation in Kenya*, 27.

48 Whitfield et al., "Conceptualising Farming Systems."

49 Monsanto, "Water Efficient Maize for Africa."

50 Monsanto, *User Guide for the Production of YieldGard®*, 2.

51 Tumusiime et al., "The Cost of Coexistence," 209.

52 Campagne et al., "Dominant Inheritance of Field," e69675.

53 Tumusiime et al., "The Cost of Coexistence," 215.

54 Ibid., 208–21. This dense arrangement of smallholder farmers is typical of agricultural systems across SSA: studies of maize farming

systems in Zambia and Ghana also depict high rates of pollen-mediated gene flow, which serve to undermine the rigid segregation needed to keep resistance at bay. As mentioned in chapter 4 the high prevalence of seed sharing, recycling, and replanting make the challenge of maintaining strict separation between GM and non-GM cultivars even more difficult. See Aheto et al., "Implications of GM Crops"; Bøhn et al., "Co-Existence Challenges"; and Bøhn et al., "Pollen-Mediated Gene Flow."

55 Interview with Informant 85, Nairobi, 31 May 2016.

56 Qaim, *Genetically Modified Crops and Agricultural Development*, 54; interview with Informant 112, phone interview, 25 May 2017.

57 Mulaa et al., "Evaluation of Stem Borer Resistance."

58 Qaim, *Genetically Modified Crops and Agricultural Development*, 70.

59 Van den Berg, "Insect Resistance Management in Bt Maize," 223.

60 Erasmus, Marais, and Van Den Berg, "Movement and Survival of Busseola Fusca," 2288. See also Murphy, Ginzel, and Krupke, "Evaluating Western Corn Rootworm"; and Onstad et al., "Seeds of Change."

61 For a comprehensive review of this research see Van den Berg, "Insect Resistance Management in Bt Maize," 224.

62 Interview with Informant 116, phone interview, 20 June 2017.

63 Campagne et al., "Impact of Violated High-Dose Refuge Assumptions," 597.

64 Qaim, *Genetically Modified Crops and Agricultural Development*; Kruger et al., "Dominant Inheritance."

65 Tabashnik, Brévault, and Carrière, "Insect Resistance to Bt Crops," 511.

66 Campagne et al., "Impact of Violated High-Dose Refuge Assumptions," 604.

67 Interview with Informant 116, phone interview, 20 June 2017, and interview with Informant 123, phone interview, 31 May 2017.

68 Demers-Morris, "GEOs and Gender," 64.

69 Interview with Informant 85, Nairobi, 31 May 2016.

70 Ariga et al., *Trends and Patterns in Fertilizer Use.*

71 See for instance Waithaka et al., "Factors Affecting the Use of Fertilizers"; Onduru et al., "Manure and Soil Fertility Management."

72 Muhunyu, "Structural Analysis of Small-Scale Maize Production"; Odhiambo et al., "Weed Dynamics During Transition," E0133976.

73 Kalaitzandonakes, Kruse, and Gouse, "The Potential Economic Impacts," 222.

74 Demers-Morris, "GEOs and Gender," 84.

75 Interview with Informant 85, Nairobi, 31 July 2015.

76 See Arends-Kuenning and Makundi, "Agricultural Biotechnology for Developing Countries"; and Schnurr, "Inventing Makhathini."

77 Interview with Informant 85, Nairobi, 31 May 2016.

78 Croppenstedt, Goldstein, and Rosas, "Gender and Agriculture"; Peterman, Behrman, and Quisumbing, *A Review of Empirical Evidence on Gender Differences.*

79 Another publication by the same research group that emerged around the same time stressed the importance of integrating gender into agricultural biotechnology programming, but the recommendations themselves are very broad (for example, greater representation of women within education programming and as extension officers, increased involvement of women farmers in trait selection, etc.). Also, there is no explicit mention of the WEMA program in this document. See Ezezika, Deadman, and Daar, "She Came, She Saw, She Sowed." WEMA's sister program – Drought Tolerant Maize for Africa (DTMA) – seems to have paid comparatively more attention to gender dynamics than WEMA: in 2012 DTMA hired a gender consultant and recommitted itself to fostering equitable participation of men and women (Kandiwa and Tegbaru, "Gender in DTMA").

80 Interview with Informant 85, Nairobi, 31 May 2016.

81 Demers-Morris, "GEOs and Gender," 75–6.

82 Interview with Informant 123, phone interview, 31 May 2017.

83 Kalule et al., "Farmers' Perceptions of Importance"; Moyin-Jesu, "Comparative Evaluation of Modified Neem Leaf"; Oben et al., "Farmers Knowledge and Perception."

84 Mihale, "Use of Indigenous Knowledge."

85 Whitfield et al., "Conceptualising Farming Systems," 54.

86 Lynas, "Tanzania Is Burning GM Corn While People Go Hungry."

87 Solomon, "Farmers Search for Answers."

88 Uys, "GM Maize Could Be Less Susceptible to Fall Armyworm"; Company officials have been reluctant to broadcast this, as Bt maize is registered to control two species of maize pests in South Africa and FAW is not one of them, which had led to seed companies filing complaints against each other with the South Africa National Seed Organization (interview with Informant 116, phone interview, 20 June 2017).

89 Sylvester Oikeh, quoted in Moseti and Odunga, "Food Disaster Imminent."

90 Wamuswa, "New Maize Variety Gives Farmers Hope."
91 Stephen Ochen, quoted in Lutaaya, "Experts Speak Out on Pesticide-Resistant Armyworms."
92 Tembo, "Genetically-Modified Bt Maize Alternative."
93 Moseti and Odunga, "Food Disaster Imminent."
94 Quote from Interview with Informant 116, phone interview, 20 June 2017. There are also cases of FAW developing resistance to Bt maize. In Brazil, FAW has shown resistance to both MON 810 and Cry1F from Pioneer, while similar reports of resistance to the Cry1F event have been reported in the southeastern United States and Puerto Rico (interview with Informant 116, phone interview, 20 June 2017; Huang et al., "Cry2F Resistance"; Storer et al., "Status of Resistance").

CHAPTER SEVEN

1 NARO, *Economic Impact Brief of Different Crops.*
2 Uganda Biosciences Information Center, *Facts and Figures.*
3 NARO, *Economic Impact Brief of Different Crops.*
4 Ibid.
5 Interview with Informant 5 and Informant 7, Kawanda Agricultural Research Institute, 23 June 2010, and 6 May 2012.
6 The specific strain prevalent in Uganda is banana xanthomonas wilt.
7 Karamura et al., "Assessing the Impacts of Banana Bacterial Wilt"; Rietveld et al., "Effect of Banana Xanthomonas Wilt."
8 Interview with Informant 104, Nairobi, 2 June 2016. See also Tushemereirwe et al., "An Outbreak," "First Report"; and Yirgou and Bradbury, "A Note on Wilt."
9 Scovia Adikini et al., "Spread of Xanthomonas Campestris."
10 Ligami, "Bacterial Wilt-Resistant Banana Variety in the Offing."
11 Dale et al., "Modifying Bananas: From Transgenics to Organics?"
12 Interview with Informant 5 and Informant 7, Kawanda Agricultural Research Institute, 30 June 2010.
13 Interview with Informant 29, Kampala, 2 June 2015.
14 Interview with Informant 51, Kampala, 9 May 2011; Schnurr, "Biotechnology and Bio-Hegemony in Uganda."
15 These agreements cover experimentation, but they do not extend to commercialization (interview with Informant 29, Kampala, 2 June 2015).
16 Interview with Informant 12, Kampala, 6 June 2009.
17 Interview with Informant 13, Namulonge Agricultural and Animal Production Research Institute, 31 May 2009.

18 Interview with Informant 51, Kampala, 12 May 2011.

19 Certain districts and sub-counties were excluded based on health and safety concerns, as well as inaccessibility due to flooding during the rainy season.

20 Bellon and Reeves, "Quantitative Analysis of Data"; Witcombe et al., "Participatory Plant Breeding Is Better"; Ceccarelli and Grando, "Decentralized-Participatory Plant Breeding"; Scoones and Thompson, *Farmer First Revisited.*

21 Focus group with male farmers, Kamuli District, 18 June 2013.

22 Focus group with female farmers, Kamuli District, 17 June 2013.

23 Focus group with female farmers, Kyenjojo District, 22 November 2013.

24 Biesalski, *Hidden Hunger.*

25 Uganda Bureau of Statistics and ICF International, "Uganda Demographic and Health Survey."

26 Jonnalagadda et al., "How Cost-Effective is Biofortification"; Van Der Straeten, Fitzpatrick, and De Steur, "Editorial Overview: Biofortification of Crops," vii–x.

27 Mbabazi, *Molecular Characterisation and Carotenoid Quantification.*

28 Schnurr, Addison, and Mujabi-Mujuzi, "Limits to Biofortification."

29 Focus group with female farmers, Wakiso District, 13 December 2012; focus group with male farmers, Ntungamo District, 15 May 2014.

30 Focus group with male farmers, Ntungamo District, 15 May 2014.

31 Interview, Informant 29, Kawanda Agricultural Research Institute, 2 June 2014.

32 Tushemereirwe et al., *An Outbreak of Banana Bacterial Wilt*; Shimwela et al., "Local and Regional Spread of Banana."

33 Karamura and Tinzaara, *Management of Banana Xanthomonas Wilt,* 27.

34 An expanded version of this summary of the scale of devastation wrought by BBW can be found in Addison and Schnurr, "Growing Burdens?"

35 The proof of concept for the BBW-resistant banana was established in 2010, using transgenic material from the pepper plant, given licence free from a Taiwanese biotech company. The first three generations were raised in confined field trials at NARO's Kawanda Agricultural Research Centre in 2012. As of 2016 breeders had developed twelve separate transgenic lines that showed 100 per cent resistance to BBW over three generations. Currently these lines are undergoing multi-locational testing across Uganda, and should be ready for commercial

release by 2020 (interview with Informant 104, Nairobi, 2 June 2016; interview with Informant 29, Kampala, 27 May 2016).

36 Jogo et al., "Determinants of Farm-Level Adoption."

37 Interview with Informant 29, Kampala, 2 June 2015.

38 Ndungo et al., "Presence of Banana Xanthomonas Wilt," 294; Reeder et al., "Presence of Banana Bacterial Wilt," 1038; Carter et al., "Identification of Xanthomonas Vasicola," 403; Mutua, "Watch Out for Banana Disease"; Shimwela et al., "Local and Regional Spread of Banana Xanthomonas Wilt."

39 Kubiriba, Erima, and Tushemereirwe, "Scaling out Control of Banana Xanthomonas Wilt."

40 NARO BBW maps provided by Drake Mubiru, research officer at NARO Kawanda (personal communication, 2 January 2018). One reason that compliance rates are so low is that all of these agronomic measures of control are extremely labour intensive: over the years some banana breeders have disclosed to me that even they do not follow through on all of these control measures because they are too laborious (interview with Informant 5, Kawanda Agricultural Research Institute, 16 May 2014). For more information on the potential labour impacts of a BBW-resistant GM variety, see Addison and Schnurr "Growing Burdens?"

41 Gibson and Hotz, "Dietary Diversification/Modification Strategies"; Nair, Augustine, and Konapur, "Food-Based Interventions," 277; Suri, Kumar, and Das, "Dietary Deficiency of Vitamin A."

42 Chandrasekharan, "Sustainable Control of Vitamin A Deficiency," 786; Talukder et al., "Increasing the Production and Consumption of Vitamin A"; Cabalda et al., "Home Gardening Is Associated"; Faber and Laurie, "A Home Gardening Approach"; Faber and Jaarsveld, "The Production of Provitamin A-Rich Vegetables."

43 Ibid.; Talukder et al., "Increasing the Production and Consumption."

44 Nair, Augustine, and Konapur, "Food-Based Interventions to Modify Diet," 277.

45 Kikafunda et al., *Nutrition Country Profile the Republic of Uganda*.

46 Interview with Informant 29, Kampala, 4 June 2014.

47 Glover, Sumberg, and Andersson, "The Adoption Problem," 3–6.

48 Interview with Informant 29, Kampala 14 December 2012. The host varieties being used are the same in both sets of GM bananas being bred to resist BBW and those biofortified with provitamin A (interview with Informant 29, Kampala, 27 May 2016).

49 Afedraru, "The Story Behind Conventional Banana Breeding."

50 NARO, *Economic Impact Brief of Different Crops.*

51 Interview with Informant 5 and Informant 7, Kawanda Agricultural Research Institute, 6 May 2012.

52 Focus group with female farmers, Wakiso District, 13 December 2012.

53 Ibid.

54 Focus group with female farmers, Kamuli District, 17 June 2013.

55 Interview with Informant 120, Kawanda Agricultural Research Centre, 11 May 2017.

56 Focus group with male farmers, Kamuli District, 18 June 2013.

57 Interview with Informant 120, Kawanda Agricultural Research Centre, 11 May 2017.

58 Nowakunda et al., "Consumer Acceptability of Introduced Bananas in Uganda."

59 Interview with Informant 109, Kampala, 10 May 2017.

60 In hindsight, banana breeders reflected that they "targeted the wrong consumers" by focusing on those who consume matooke as their primary staple, and should have instead focused on growers in these outlying regions who are less fickle about the various value-based characteristics such as taste, colour, and texture (interview with Informant 109, Kampala, 10 May 2017). As the former head of the banana breeding program explained in a local broadsheet: "Those who say that it is not tasty compare it to the local banana varieties, but to a person who has never eaten the local varieties, Fhia is quite tasty" (Muzaale,"Small Market").

61 Focus group with male farmers, Kamuli District, 18 June 2013.

62 Focus group with female farmers, Buikwe District, 10 December 2012.

63 Focus group with female farmers, Wakiso District, 13 December 2012.

64 Interview with Informant 29, Kawanda Agricultural Research Centre, 27 May 2016.

65 Kabunga, Dubois, and Qaim, "Impact of Tissue Culture Banana Technology."

66 The onset of resistance occurs via two modalities: sexual reproduction leads to new recombination of genes and/or the pathogen evolves new individuals that no longer respond to the resistance mechanism. Because banana reproduces clonally, the appearance of the first modality is basically impossible, so breeders are focused exclusively on the second. The pace of resistance will be determined largely by the pathogen's reproductive rates – that is, a fungus, which reproduces both sexually and asexually, will develop resistance faster than a bacterium, which only reproduces asexually. As such, a GM banana resistant to

black sigatoka, a fungus, is expected to need to be replaced sooner than a GM banana resistant to BBW, a bacterium. Breeders are stacking transgenic material to delay resistance as long as possible, but the development of resistance is "inevitable" (interview with Informant 29, Kampala, 2 June 2015).

67 These were shown to the farmer as variety X and variety Y, to ensure that decisions were based solely on their phenotypic traits and not the method of breeding used.

68 Mean values for farmer preference were assessed on a five-point Likert scale, with "five" representing "very likely to adopt" and "one" representing "not very likely to adopt." The mean for the traditional variety was 4.39, while the mean for the hypothetical GM variety was 2.21 (standard deviations were 1.09 and 1.27, respectively). A paired t-test was undertaken, and the preference for longer-standing varieties was statistically significant ($t = 12.57$, $p < .001$).

69 Focus group with female farmers, Wakiso District, 13 December 2012.

70 Focus group with male farmers, Buikwe District, 11 December 2012.

71 Focus group with female farmers, Buikwe District, 10 December 2012.

72 Focus group with male farmers, Nakaseke District, 12 December 2012.

73 Soleri et al., "Testing Assumptions Underlying Economic Research," 669.

74 Mean values for farmer preference were assessed on a five-point Likert scale, with "five" representing "very likely to adopt" and "one" representing "not very likely to adopt." The mean for the traditional variety was 4.63, while the mean for the hypothetical GM variety was 1.71 (standard deviations were 0.85 and 0.90, respectively). A paired t-test was undertaken, and the preference for longer-standing varieties was statistically significant ($t = 24.94$, $p < .001$). Note that "variety X" provides medium-level yields over a six-year period, while "variety Y" has high yields initially, followed by medium yields after three years, and low yields after six years.

75 Focus group with male farmers, Buikwe District, 11 December 2012.

76 Focus group with male farmers, Nakaseke District, 12 December 2012.

77 Focus group with female farmers, Buikwe District, 10 December 2012.

78 Mucioki et al., "Supporting Farmer Participation in Formal Seed Systems."

79 Abdi and Nishikawa, "Understanding Smallholder Farmers' Access."

80 Interview with Informant 114, Kawanda Agricultural Research Institute, 12 May 2017.

81 Rietveld, Ajambo and Kikulwe, "Economic Gain and Other Losses."
82 For more on the potential gendered impacts of a GM banana, see Addison and Schnurr, "Growing Burdens?" For more on the gendered dynamics of banana production in Uganda, see Beraho, *Living with AIDS*.
83 NARO breeders earn less than US$500 per month, which is a lot by Ugandan standards but only a fraction of the salaries paid to contemporaries with equal training who are employed by multinational corporations or the UN. In order to supplement their incomes, some breeders have started or joined private agro-dealers, and are positioned to profit considerably from the delivery system should GM matooke be commercialized. Others remain hopeful that as co-owners of the intellectual property they might strike it rich if the GM constructs designed for matooke are transplanted into the Cavendish banana, which is eaten all over the world (interview with Informant 29, Kampala, 5 May 2015).
84 Interview with Informant 29, Kampala, 2 June 2015.
85 Kikulwe, "Banana Tissue Culture."
86 Addison and Schnurr, "Growing Burdens?"
87 Interview with Informant 5 and Informant 7, Kawanda Agricultural Research Institute, 23 June 2010.

CONCLUSION

1 Smale, "GMOs and Poverty," 145.
2 Stone, "Constructing Facts: Bt Cotton in India." See also Smale, "GMOs and Poverty."
3 Stone, "Constructing Facts: Bt Cotton in India," 67.
4 Schnurr and Addison, "Which Variables Influence Farmer Adoption."
5 Addison and Schnurr, "Growing Burdens."
6 Falck-Zapeda and Zambrano, *Gender Impacts of Genetically Engineered Crops*; Gouse et al., "Genetically Modified Maize."
7 Smale, Heisey, and Leathers, "Maize of the Ancestors and Modern Varieties"; Lunduka, Fisher, and Snapp, "Could Farmer Interest in a Diversity of Seed Attributes"; De Groote et al., "Maize for Food and Feed in East Africa"; Soleri et al., "Testing Assumptions Underlying Economic Research"; Glover, "The System of Rice Intensification."
8 See Jacobson, *From Betterment to Bt Maize*, 121.
9 Andersson and Sumberg, "Knowledge Politics in Development-Oriented Agronomy," 6.

10 Wambugu, "Why Africa Needs Agricultural Biotech," 16.
11 Wambugu, "The Importance of Political Will," 3.
12 Interview with Informant 18, Nairobi, 10 May 2010.
13 Anthony and Ferroni, "Agricultural Biotechnology and Smallholder Farmers," 279.
14 Interview with Informant 118, phone interview, 15 August 2017.
15 Feed the Future, *U.S. Government's Global Food Security Strategy*; World Bank, *Unlocking Africa's Agricultural Potential*; USAID, "Agriculture and Food Security"; interview with Informant 115, Nairobi, 14 May 2017.
16 Sulle and Hall, "Reframing the New Alliance Agenda."
17 Brooks et al., "Silver Bullets, Grand Challenges and the New Philanthropy."
18 Andersson and Sumberg, "Knowledge Politics in Development-Oriented Agronomy," 5.
19 Whitfield et al., "Conceptualising Farming Systems," 60.
20 The reductive tendencies of genetic modification are most pronounced in maize breeding, where limitations around trait segregation made the creation of genetically modified versions of OPVs impossible. The obstacles that scuttled the IRMA project – particularly the tension between the management regime prescribed by the genetically modified construct and the growing habits preferred by farmers – underline this foundational incompatibility (Jacobson, *From Betterment to Bt*, 59).
21 Whitfield et al., "Conceptualising Farming Systems," 61.
22 Dibden, Gibbs, and Cocklin, "Framing GM Crops as a Food Security Solution," 60.
23 Stone, "Constructing Facts: Bt Cotton in India."
24 Ssonko, "ActionAid Admits to Misleading Ugandans"; ActionAid UK, "Response to Criticism of Actionaid."
25 Paarlberg, "A Dubious Success," 224; See also Lynas, "Lecture to Oxford Farming Conference"; Lynas, "Time to Call Out the Anti-GMO Conspiracy Theory"; and Zhu, "Calestous Juma on Being Pro-Africa."
26 Stone, "Constructing Facts: Bt Cotton in India."
27 Conrow, "Burkina Faso Puts GM Cotton on Hold."
28 GM Watch, "Burkina Faso Abandons GM Bt Cotton."
29 National Academies of Sciences, Engineering, and Medicine, *Genetically Engineered Crops*; European Food Safety Authority, "GMO."
30 Sumberg, "Could the 3Ds Breathe New Life into Farming Systems Research?"

31 Brooks and Loevinsohn, "Shaping Agricultural Innovation Systems"; Crane, "Bringing Science and Technology Studies into Agricultural Anthropology."

32 Whitfield, *Adapting to Climate Uncertainty in African Agriculture*, 132.

33 Andersson and Sumberg, "Knowledge Politics in Development-Oriented Agronomy."

34 Giller et al., "Beyond Conservation Agriculture," 870.

35 Abate et al., "Factors that Transformed Maize Productivity in Ethiopia."

36 Glover, Venot, and Maat, "On the Movement of Agricultural Technologies."

37 CABI, "Plant Clinics."

38 Of course, in order to enact transformational change, such micro-level initiatives would need to be supported by broader political and economic reforms of the global food system, including prioritizing food as a public good; enabling the state to play a larger role in supporting small-scale farming via subsidizing access to seed, inputs, and credit; limiting free markets for land; and empowering international redistributive global institutions that could regulate food prices and ensure improved access for marginalized actors like smallholder farmers (Akram-Lodhi, *Hungry for Change*).

39 Dowd-Uribe, "Engineering Yields and Inequality?" 169.

40 Altieri, "Agroecology, Small Farms, and Food Sovereignty," 103; Wezel et al., "Agroecology as a Science, a Movement or a Practice."

41 McIntyre et al. "Agriculture at a Crossroads," 3; UNHRC, "Report Submitted by the Special Rapporteur."

42 Lacombe, Couix, and Hazard, "Designing Agroecological Farming Systems with Farmers," 208.

43 Hockin-Grant and Yasué, "The Effectiveness of a Permaculture Education."

44 Bezner Kerr et al., "Participatory Research on Legume Diversification"; Kangmennaang et al., "Impact of a Participatory Agroecological Development."

45 Oakland Institute, "Agroecology Case Studies."

46 Vieira, "Community-Led Approach."

47 McKnight Foundation, "Farmer Research Networks," 1; Nelson, Coe, and Haussmann, "Farmer Research Networks."

48 Scoones and Thompson, "The Politics of Seed in Africa's Green Revolution."

49 Stone, "Agricultural Deskilling"; Stone, "Towards a General Theory"; Glover, Venot, and Maat, "On the Movement of Agricultural Technologies."

50 De Schnutter and Vanloqueren, "The New Green Revolution," 3.

51 Vanloqueren and Baret, "How Agricultural Research Systems Shape"; Lescourret, "Agroecological Engineering."

52 Ronald, "Lab to Farm"; Cyranoski, "CRISPR Tweak May Help"; Xiong et al., "Genome-Editing Technologies"; Khatodia et al., "The CRISPR/Cas Genome-Editing."

53 Li et al., "High-Efficiency TALEN-Based Gene Editing"; Ongu, "Does Uganda Need GMOs?"; Fleming, "Science's Search for a Super Banana."

54 Voytas, Gao, and McCouch, "Precision Genome Engineering."

55 See, for instance, Ferguson, *The Anti-Politics Machine*; Tilley, *Africa as a Living Laboratory*; Van Beusekom, *Negotiating Development*; Mitchell, *Rule of Experts Egypt, Techno-Politics, Modernity*; and Isaacman and Roberts, *Cotton, Colonialism, and Social History in Sub-Saharan Africa.*

56 Hodge, *Triumph of the Expert*, 8.

57 Ibid., 267.

58 Ibid.,19.

59 Harsh and Smith, "Technology, Governance and Place."

60 Hodge, *Triumph of the Expert*, 3.

61 Giller et al., "A Golden Age for Agronomy?"

62 Interview with Informant 13, Namulonge Agricultural and Animal Production Research Institute, 8 June 2009.

63 Interview with Informant 114, Kawanda Agricultural Research Institute, 12 May 2017.

64 Giller et al., "A Golden Age for Agronomy?"

Bibliography

AATF (African Agricultural Technology Foundation). "About Us." Last
    modified 2012. http://www.AATF-africa.org/about-us.
– "Efforts Underway to Develop Maize Varieties Tolerant to Drought,
    Pests." Last modified 2012. http://wema.AATF-africa.org/news/media/
    efforts-underway-develop-maize-varieties-tolerant-drought-pests.
– "Project Brief." Last modified 2012. https://wema.AATF-africa.org/
    project-brief.
Abate, Tsedeke, Bekele Shiferaw, Abebe Menkir, Dagne Wegary, Yilma
    Kebede, Kindie Tesfaye, Menale Kassie, Gezahegn Bogale, Berhanu
    Tadesse, and Tolera Keno. "Factors that Transformed Maize
    Productivity in Ethiopia." *Food Security* 7, no. 5 (2015): 965–81. http://
    doi.org/10.1007/s12571-015-0488-z.
Abate, Tsedeke, Monica Fisher, Girma Kassie, Rodney Lunduka, Paswel
    Marenya, and Woinishet Asnake. "Characteristics of Maize Cultivars in
    Africa: How Modern Are They and How Many Do Smallholder
    Farmers Grow?" *Agriculture and Food Security* 6, no. 30 (2017): http://
    doi.org/10.1186/s40066-017-0108-6.
Abdallah, Naglaa A. "GM Crops in Africa: Challenges in Egypt." *GM
    Crops and Food* 1, no. 3 (2010): 116–19. http://dx.doi.org/10.4161/
    gmcr.1.3.12811.
– "Biotech Crops and the Egyptian Revolution: Where We Stand." *GM
    Crops and Food* 2 no. 2 (2011): 83–4. http://dx.doi.org/10.4161/
    gmcr.2.2.17360.
Abdeldaffie, Elhadi, E.A. Elhag, and N.H.H. Bashir. "Resistance in the
    Cotton Whitefly, *Bemisia tabaci* (Genn.), to Insecticide Recently
    Introduced into Sudan Gezira." *Tropical Pest Management* 33 (1987):
    283–6. http://doi.org/10.1080/09670878709371171.

Abdi, Bedru Beshir, and Yoshiaki Nishikawa. "Understanding Smallholder Farmers' Access to Maize Seed and Seed Quality in the Drought-Prone Central Rift Valley of Ethiopia." *Journal of Crop Improvement* 31, no. 3 (2017): 289–310. http://doi.org/10.1080/15427528.2017.1302031.

Abdo, Eman M., Omar M. Barbary, and Omayma El-Sayed Shaltout. "Chemical Analysis of Bt Corn 'Mon-810: Ajeeb-YG' and Its Counterpart Non-Bt Corn 'Ajeeb.'" *Journal of Applied Chemistry* 4, no. 1 (2013): 55–60.

– "Feeding Study with Bt Corn (Mon 810: Ajeeb YG) on Rats: Biotechmical Analysis and Liver Histopathology." *Food and Nutrition Sciences* 5 (2014): 185–95. http://doi.org/10.4236/fns.2014.52024.

ABNE (African Biosafety Network of Expertise). "ABNE Brochure." N.d. http://nepad-abne.net/about-us/abne-brochure.

– "About Us." Last modified 2016. http://nepad-abne.net/about-us.

– "Supporting Malawi in Overall Regulatory Handling of Bt Cotton Multi-Location Trials." *ABNE Newsletter*, April–June 2014. http://nepad-abne.net/wp-content/uploads/2015/11/ABNE Newsletter-April-to-June-reports_Final.pdf 2014.

– *ABNE in Africa: Building Functional Biosafety Systems in Africa.* Ouagadougou, Burkina Faso, 2014: ANBE. http://nepad-abne.net/wp-content/uploads/2015/11/ABNE-in-Africa-with-cover-1-28-13.pdf

Abrahams, Zulfa, Zandile Mchiza, and Nelia P. Steyn. "Diet and Mortality Rates in Sub-Saharan Africa: Stages in the Nutrition Transition." *BMC Public Health* 11 (2011): 801. http://doi.org/10.1186/1471-2458-11-801.

ABSP II (Agricultural Biotechnology Support Project II). "What Is ABSPII?" Last modified 2013. http://www.absp2.cornell.edu/aboutabsp2.

– "Agricultural Biotechnology Support Project II: Supporting Agricultural Development through Biotechnology," 2010. http://www.absp2.cornell.edu.

ACB (African Centre for Biodiversity). *Africa Bullied to Grow Defective Bt Maize: The Failure of Monsanto's MON810 Maize in South Africa.* South Africa: African Centre for Biodiversity, 2013.

– *Profiting from the Climate Crisis, Undermining Resilience in Africa: Gates and Monsanto's Water Efficient Maize for Africa (WEMA) Project.* South Africa: African Centre for Biodiversity, 2015.

ACB (African Centre for Biosafety) and AFSA (Alliance for Food Sovereignty in Africa). "Modernising African Agriculture: Who Benefits?" GRAIN, 17 May 2013. https://www.grain.org/fr/bulletin_board/entries/4727-modernising-african-agriculture-who-benefits.

ActionAid U K. "Response to Criticism of Actionaid Uganda's Genetically Modified Foods Campaign." ActionAid, 8 June 2015. https://www.actionaid.org.uk/latest-news/response-to-criticism-of-actionaid-ugandas-campaign.

Addison, Lincoln, and Matthew Schnurr. "Growing Burdens? Disease-Resistant Genetically Modified Bananas and the Potential Gendered Implications for Labor in Uganda." *Agriculture and Human Values* 33, no. 4 (2016): 967–78. http://doi.org/10.1007/s10460-015-9655-2.

Adenle, Ademole A., Jane E. Morris, and Govindan Parayil. "Status of Development, Regulation and Adoption of G M Agriculture in Africa: Views and Positions of Stakeholder Groups." *Food Policy* 43 (2013): 159–66. http://doi.org/10.1016/j.foodpol.2013.09.006.

Adikini, Scovia, Fen Beed, Geoffrey Tusiime, Leena Tripathi, Samuel Kyamanywa, Melanie Lewis-Ivey, and Sally A. Miller. "Spread of *Xanthomonas campestris* pv. *musacearum* in Banana Plants: Implications for Management of Banana Xanthomonas Wilt Disease." *Canadian Journal of Plant Pathology* 35, no. 4 (2013): 458–68. http://doi.org/10.1080/07060661.2013.845856.

Afedraru, Lominda. "The Story Behind Conventional Banana Breeding to Fight Pests and Diseases." *Daily Monitor*, 12 December 2012. http://www.monitor.co.ug/Magazines/Farming/Behind-conventional-banana-breeding-to-fight-pests-and-diseases/689860-1641826-fq7uv6z/index.html.

Afedraru, Lominda, and Robert Kafeero Sekitoleko. "Ugandan Legislator on Why His Nation Needs to 'Wake Up' on G M O s" *Genetic Literacy Project*, 27 April 2018, https://geneticliteracyproject.org/2018/04/27/ugandan-legislator-on-why-his-nation-needs-to-wake-up-on-gmos.

Agaba, John. "Ugandan Scientists Skeptical of Revised G M O Bill." *Cornell Alliance for Science*, 29 November 2018. https://allianceforscience.cornell.edu/blog/2018/11/ugandan-scientists-skeptical-revised-gmo-bill.

A G R A (Alliance for a Green Revolution in Africa). "Our Story." N.d. https://agra.org/who-we-are.

– "Program Development and Innovation." N.d. https://agra.org/program-development-and-innovation.

– "Frequently Asked Questions (F A Q s)." Last modified 2017. https://agra.org/faqs/#1500366462377-91e54a88-c92d.

– *African Agriculture Status Report: Focus on Staple Crops*. Nairobi, Kenya: Alliance for a Green Revolution in Africa, 2013.

– "A G R A Focus," 2016. https://agra.org/2016AnnualReport/agra-focus-2.

- *Annual Report: Towards Africa's Agricultural Transformation. Kenya: Alliance for a Green Revolution in Africa.* Nairobi, Kenya: Alliance for a Green Revolution in Africa, 2016. https://agra.org/annual-reports.
- *Strategy Overview for 2017–2021: Inclusive Agricultural Transformation in Africa.* Nairobi, Kenya: Alliance for a Green Revolution in Africa, 2016. https://agra.org/wp-content/uploads/2018/02/AGRA-Corporate-Strategy-Doc-3.-2.pdf.

Agyeman, Nana Konadu. "Court Dismisses Injunction on Commercialising of GMOs." *Graphic Online*, 30 October 2015. https://www.graphic.com.gh/news/general-news/court-dismisses-injunction-on-commercialising-of-gmos.html.

Aheto, Denis W., Thomas Bøhn, Broder Breckling, Johnnie van den Berg, Lim Lee Ching, and Odd-Gunner Wikmark. "Implications of GM Crops in Subsistence-Based Agricultural Systems in Africa." In *GM-Crop Cultivation: Ecological Effects on a Landscape Scale*, edited by Broder Breckling and R. Verhoeven: 93–103. Frankfurt: Peter Lang, 2013.

Akram-Lodhi, A. Haroon. *Hungry for Change: Farmers, Food Justice and the Agrarian Question.* Halifax, NS: Fernwood Publishing, 2013.

Anderson, Kym, Ernesto Valenzuela, and Leeann Jackson. "Recent and Prospective Adoption of Genetically Modified Cotton: A Global Computable General Equilibrium Analysis of Economic Impacts." *Economic Development and Cultural Change* 56, no. 2 (2008): 265–96. http://doi.org/10.1086/522897.

Andersson, Jens A., and James Sumberg. "Knowledge Politics in Development-Oriented Agronomy." In *Agronomy for Development: The Politics of Knowledge in Agricultural Research*, edited by James Sumberg. New York: Routledge, 2017.

Anthony, Vivenne, and Marco Ferroni. "Agricultural Biotechnology and Smallholder Farmers in Developing Countries." *Current Opinion in Biotechnology* 23, no. 2 (2012): 278–85. http://doi.org/10.1016/j.copbio.2011.11.020.

Arends-Kuenning, Mary, and Flora Makundi. "Agricultural Biotechnology for Developing Countries. Prospects and Policies." *American Behavioral Scientist* 44, no. 3 (2000): 318–49. http://doi.org/10.1177/0002764 0021956242.

Ariga, Joshua, T.S. Jayne, Betty Kibaara, and J.K. Nyoro. *Trends and Patterns in Fertilizer Use by Smallholder Farmers in Kenya, 1997–2007.* Nairobi, Kenya: Tegemeo Institute of Agricultural Policy and Development, 2008.

Assefa, Kebebew. "The Dire Need to Support 'Orphan Crop' Research." *Appropriate Technology* 41, no. 2 (2014): 8–9. ISSN: 03050920.

Assefa, Yoseph, and Johnnie Van den Berg. "Genetically Modified Maize: Adoption Practices of Small-Scale Farmers in South Africa and Implications for Resource Poor Farmers on the Continent." *Aspects of Applied Biology* 96 (2009): 215–23.

Assem, Shireen, K. "Opportunities and Challenges of Commercializing Biotech Products in Egypt: Bt Maize: A Case Study." In *Biotechnology in Africa: Emergence Initiatives, Future,* edited by Florence Wambuga, Daniel Kamanga, and Ismail Serageldin, 37–52. Switzerland: Springer, 2013.

AU (African Union). *African Model Law on Safety in Biotechnology*, 2001.

Ayele, Seife. "The Legitimation of GMO Governance in Africa." *Science and Public Policy* 34, no. 4 (2007): 239–49. http://doi.org/10.3152/030234207X213931.

Ayele, Seife, and David Wield. "Science and Technology Capacity Building and Partnership In African Agriculture: Perspectives on Mali and Egypt." *Journal of International Development* 17 (2005): 631–46. http://doi.org/10.1002/jid.1228.

Badiane, Ousmane, and Tsitsi Makombe. "The Theory and Practice of Agriculture, Growth, and Development in Africa." WIDER working paper 2014/061, World Institute for Development Economics Research at the United Nations University (UNU-WIDER), Tokyo, Japan, 2014. https://www.wider.unu.edu/sites/default/files/wp2014-061.pdf.

Baffes, John. "The 'Full Potential' of Uganda's Cotton Industry." *Development Policy Review* 27, no. 1 (2009): 67–85. http://doi.org/10.1111/j.1467-7679.2009.00436.x.

– "The Cotton Sector of Uganda." Africa Region working paper series no. 123, World Bank, 2009: 1–30.

Balch, Oliver. "Are We Able to Have a Rational Debate about GM?" *Guardian*, 26 September 2013.

Bassett, Thomas J. *The Peasant Cotton: Revolution in West Africa, Côte D'Ivoire, 1880–1995*. Cambridge, NY: Cambridge University Press, 2001.

Bavier, Joe. "Burkina Faso Settles Dispute with Monsanto over GM Cotton." *Reuters*, 8 March 2017.

– "How Monsanto's GM Cotton Sowed Trouble in Africa." *Reuters*, 8 December 2017.

Bellon, Mauricio R., and Jane Reeves. "Quantitative Analysis of Data From Participatory Methods In Plant Breeding." Manuals 23718, International Maize and Wheat Improvement Center (CIMMYT), 2002. https://ideas.repec.org/p/ags/cimmma/23718.html.

Bennett, Richard, Stephen Morse, and Yousouf Ismael. "The Economic Impact of Genetically Modified Cotton on South African Smallholders: Yield, Profit and Health Effects." *Journal of Development Studies* 42, no. 4 (2007): 662–77. http://doi.org/10.1080/00220380600 682215.

Beraho, Monica K. *Living with AIDS in Uganda: Impacts on Banana-Farming Households in Two Districts*. Wageningen, Netherlands: Wageningen Academic Publishers, 2008.

Bernal, Victoria. "Cotton and Colonial Order in Sudan: A Social History, with Emphasis on the Gezira Scheme." In *Cotton, Colonialism, and Social History in Sub-Saharan Africa*, edited by Allen Isaacman and Richard Roberts, 96–118. Portsmouth, NH: Heinemann, 1995.

Bernstein, Henry. "Considering Africa's Agrarian Questions." *Historical Materialism* 12, no. 4 (2004): 115–44. http://doi.org/10.1163/1569206043505158.

Bezner Kerr, Rachel, Sieglinde Snapp, Marko Chirwa, Lizzie Shumba, and Rodgers Msachi. "Participatory Research on Legume Diversification with Malawian Smallholder Farmers for Improved Human Nutrition and Soil Fertility." *Experimental Agriculture* 43, no. 4 (2007): 437–53. http://doi.org/10.1017/S0014479707005339.

Biesalski, Hans Konrad. *Hidden Hunger*. Translated by Patrick O'Mealy. Berlin, Germany: Springer, 2013.

Bimpong, Manneh, Baboucarr Manneha, Mamadou Sock, Faty Diaw, Nana Kofi Abaka Amoah, Abdelbagi M. Ismail, Glenn Gregorio, Rakesh Kumar Singh, and Marco Wopereisc. "Improving Salt Tolerance of Lowland Rice Cultivar 'Rassi' through Marker-Aided Backcross Breeding in West Africa." *Plant Science* 242 (2016): 288–99. http://doi.org/10.1016/j.plantsci.2015.09.020.

Bingen, Jim. "Cotton in West Africa: A Question of Quality." In *Agricultural Standards: The Shape of the Global Food and Fibre System*, edited by Jim Bingen and Lawerence Busch, 219–42. Dordrecht, Netherlands: Springer, 2006.

Binimelis, Rosa, Walter Pengue, and Iliana Monterroso. "'Transgenic Treadmill': Responses to the Emergence and Spread of Glyphosate-Resistant Johnsongrass in Argentina." *Geoforum* 40, no. 4 (2009): 623–33. http://doi.org/10.1016/j.geoforum.2009.03.009.

Bishop, Matthew, and Michael Green. *Philanthro-Capitalism: How Giving Can Save the World*. New York: Bloomsbury, 2009.

Blein, Roger, Martin Bwalya, Sloans Chimatiro, Benoît Faivre-Dupaigre, Simon Kisira, Henri Leturque, and Augustin Wambo-Yamdjeu.

Agriculture in Africa: Transformation and Outlook. New Partnership for African Development, 2013.

Blum, Abraham. "Effective Use of Water (EUW) and Not Water-Use Efficiency (WUE) Is the Target of Crop Yield Improvement under Drought Stress." *Field Crops Research* 112, no. 2 (2009): 119–23. http://doi.org/10.1016/j.fcr.2009.03.009.

BMGF (Bill and Melinda Gates Foundation). "Alliance for a Green Revolution in Africa: How We Work." N.d. https://www.gatesfoundation. org/How-We-Work/Resources/Grantee-Profiles/Grantee-Profile-Alliance-for-a-Green-Revolution-in-Africa-AGRA.

– "Awarded Grants." N.d. http://www.gatesfoundation.org/How-We-Work/Quick-Links/Grants-Database#q/k=maize.

– "2015 Gates Annual Letter: Our Big Bet for the Future," 2015. http://www.gatesnotes.com/2015-annual-letter.

– "Bill Gates: GMOs Will End Starvation in Africa." *Wall Street Journal*, 22 January 2016.

Boadi, Richard Y., and Mpoko Bokanga. "The African Agricultural Technology Foundation Approach to IP Management." In *Intellectual Property Management in Health and Agricultural Innovation: A Handbook of Best Practices*, edited byAnatole Krattiger et al., 1,765–74. Oxford, UK: MIHR, and Davis, CA: PIPRA, 2007.

Bøhn, Thomas, Denis W. Aheto, Felix S. Mwangala, Klara Fischer, Inger Louise Bones, Christopher Simoloka, Ireen Mbeule, Gunther Schmidt, and Broder Breckling. "Pollen-Mediated Gene Flow and Seed Exchange in Small-Scale Zambian Maize Farming, Implications for Biosafety Assessment." *Scientific Reports* 6 (2016): 34,483. http://doi.org/10.1038/srep34483.

Bøhn, Thomas, Denis W. Aheto, Felix S. Mwangala, Inger Louise Bones, Christopher Simoloka, Ireen Mbeule, and Ignacio Chapela. "Co-Existence Challenges in Small-Scale Farming When Farmers Share and Save Seeds." GM-*Crop Cultivation: Ecological Effects on a Landscape Scale*, edited by Broder Breckling and R. Verhoeven, 104–9. Frankfurt: Peter Lang, 2013.

Bortesi, Luisa, and Rainer Fischer. "The CRISPR/Cas9 System for Plant Genome Editing and Beyond." *Biotechnology Advances* 33, no. 1 (2015): 41–52. http://doi.org/10.1016/j.biotechadv.2014.12.006.

Bouët, Antoine, and Guillaume P. Gruère. "Refining Opportunity Cost Estimates of Not Adopting GM Cotton: An Application in Seven Sub-Saharan African Countries." *Applied Economic Perspectives and Policy* 33, no. 2 (2011): 260–79. http://doi.org/10.1093/aepp/ppr010.

Bowman, Andrew. "Sovereignty, Risk and Biotechnology: Zambia's 2002
   GM Controversy in Retrospect." *Development and Change* 46, no. 6
   (2015): 1,369–91. http://doi.org/10.1111/dech.12196.
Boyce, James K. *The Philippines: The Political Economy of Growth and
   Impoverishment in the Marcos Era.* Honolulu, HI: University of Hawaii
   Press, 1993.
Brenner, Carline. *Telling Transgenic Technology Tales: Lessons from the
   Agricultural Biotechnology Support Project (ABSP) Experience.* Ithaca,
   NY: International Service for the Acquisition of Agri-Biotech
   Applications (ISAAA), 2004.
Brooks, Sally. "Biofortification: Lessons from the Golden Rice Project."
   Food Chain 3, no. 1 (2013): 77–88. http://doi.org/10.3362/2046-1887.
   2013.007.
Brooks, Sally, Melissa Leach, Henry Lucas, and Erik Millstone. "Silver
   Bullets, Grand Challenges and the New Philanthropy." STEPS working
   paper 24, Brighton: STEPS Centre, 2009.
Brooks, Sally, and Michael Loevinsohn. "Shaping Agricultural Innovation
   Systems Responsive to Food Insecurity and Climate Change." *Natural
   Resources Forum* 35, no. 3 (2011): 185–200. http://doi.org/10.1111/
   j.1477-8947.2011.01396.x.
Brooks, Sally, John Thompson, Hannington Odame, Betty Kibaara, Serah
   Nderitu, Francis Karin, and Erik Millstone. *Environmental Change and
   Maize Innovation in Kenya: Exploring Pathways In and Out of Maize.*
   STEPS working paper 36, Brighton: STEPS Centre, 2009.
Bürgi, Jürg. *Insect-Resistant Maize: A Case Study of Fighting the African
   Stem Borer.* Cambridge, MA: Cabi, 2009.
Business Daily. "Uhuru Banks on Biotech Cotton for 50,000 Jobs,"
   24 January 2018. https://www.businessdailyafrica.com/economy/Uhuru-
   banks-on-biotech-cotton-for-50-000-jobs/3946234-4277552-qiuvtkz/
   index.html.
Byerlee, Derek. "Modern Varieties, Productivity, and Sustainability: Recent
   Experience and Emerging Challenges." *World Development* 24, no. 4
   (1996): 697–718. http://doi.org/10.1016/0305-750X(95)00162-6.
Byerlee, Derek, and Paul W. Heisey. "Past and Potential Impacts of Maize
   Research in Sub-Saharan Africa: A Critical Assessment." *Food Policy* 21,
   no. 3 (1996): 255–77. http://doi.org/10.1016/0306-9192(95)00076-3.
Cabalda, Aegina B., Pura Rayco-Solon, Juan Antonio A. Solon, and
   Florentino S. Solon. "Home Gardening Is Associated with Filipino
   Preschool Children's Dietary Diversity." *Journal of the American*

Dietetic Association 111, no. 5 (2011): 711–15. http://doi.org/10.1016/
j.jada.2011.02.005.

Cabanilla, Liborio S., Tahirou Abdoulaye, and John H Sanders. "Economic
Cost of Non-Adoption of Bt-Cotton in West Africa." *International
Journal of Biotechnology* 7, no. 1–3 (2005): 46–61.

CABI (Centre for Agriculture and Bioscience International). "Plant
Clinics." N.d. https://www.cabi.org/projects/our-plantwise-programme/
plant-clinics.

Cafer, Anne, and J. Sanford Rikoon. "Coerced Agricultural Modernization:
A Political Ecology Perspective of Agricultural Input Packages in South
Wollo, Ethiopia." *Journal of Rural Social Sciences* 32, no. 1 (2017):
77–97.

Campagne, Pascal, Marlene Kruger, Rémy Pasquet, Bruno Le Ru, and
Johnnie Van Den Berg. "Dominant Inheritance of Field-Evolved
Resistance to Bt Corn in Busseola Fusca." *PLoS One* 8, no. 7 (2013):
e69675. http://doi.org/10.1371/journal.pone.0069675.

Campagne, Pascal, Peter E. Smouse, Rémy Pasquet, Jean-François Silvain,
Bruno Le Ru, and Johnnie Van Den Berg. 2015. "Impact of Violated
High-Dose Refuge Assumptions on Evolution of Bt Resistance."
Evolutionary Applications 9, no. 4: 596–607. http://doi.org/10.1111/
eva.12355.

Carney, Judith A. "Converting the Wetlands, Engendering the
Environment: The Intersection of Gender with Agrarian Change in
Gambia." In *Liberation Ecologies: Environment, Development and
Social Movement*, edited by Richard Peet and Micheal Watts, 165–87.
New York: Routledge, 2002.

Carter, Brian Anthony, R. Reeder, S.R. Mgenzi, Z.M. Kinyua, J.N. Mbaka,
K. Doyle, V. Nakato, M. Mwangi, F. Beed, V. Aritua, M.L. Lewis Ivey,
S.A. Miller, and J.J. Smith. "Identification of *Xanthomonas vasicola*
(formerly *X. campestris* pv. *musacearum*), Causative Organism of
Banana Xanthomonas Wilt, in Tanzania, Kenya and Burundi." *Plant
Pathology* 59, no. 2 (2010): 403. http://doi.org/10.1111/j.1365-3059.
2009.02124.x.

Casassus, Barbara. "Paper Claiming GM Link with Tumours Republished."
Nature News, 24 June 2014. http://www.nature.com/news/paper-
claiming-gm-link-with-tumours-republished-1.15463.

Castiglioni, Paolo, Dave Warner, Robert J. Bensen, Don C. Anstrom, Jay
Harrison, Martin Stoecker, Mark Abad, Ganesh Kumar, Sara Salvador,
Robert D'Ordine, Santiago Navarro, Stephanie Back, Mary Fernandes,

Jayaprakash Targolli, Santanu Dasgupta, Christopher Bonin, Michael H. Luethy, and Jacqueline E. Heard. "Bacterial RNA Chaperones Confer Abiotic Stress Tolerance in Plants and Improved Grain Yield in Maize Under Water-Limited Conditions." *Plant Physiol* 147 (2008): 446–55. eISSN: 1532-2548.

CBC Egypt. "#Zayelshams - يز ‏سمشلا - ‏روذبلا ‏ةلدعملا ‏ايثارو ‏ةيف ‏اهيف ‏امس ‏لتاق." YouTube video, 34:40, 26 May 2013. https://www.youtube.com/watch?v=BDcRnv_ydXE.

Ceccarelli, Salvatore, and Stefania Grando. "Decentralized-Participatory Plant Breeding: An Example of Demand Driven Research." *Euphytica*, 155, no. 3 (2007): 349–60. http://doi.org/10.1007/s10681-006-9336-8.

Chambers, Judith A., Patricia Zambrano, José Benjamin Falck-Zepeda, Guillaume P. Gruère, Debdatta Sengupta, and Karen Hokanson. *GM Agricultural Technologies for Africa: A State of Affairs.* Washington, DC: International Food Policy Research Institute and African Development Bank, 2014.

Chandrasekharan, Nirmala. "Sustainable Control of Vitamin A Deficiency." *BMJ* 321 (2000): 786.

CIMMYT (International Maize and Wheat Improvement Center). "Honing Skills in Scientific Writing for Publishing." *Informa*, no. 1,792 (2012): 19–24.

– "High Expectations among Stakeholders as WEMA Phase II Kicks Off." Last modified 27 February 2013. http://www.cimmyt.org/high-expectations-among-stakeholders-as-wema-phase-ii-kicks-off.

Cipriani, Giselle, S. Fuentes, V. Bello, L.F. Salazar, M. Ghislain, and D.P. Zhang. "Transgene Expression of Rice Cysteine Proteinase Inhibitors for the Development of Resistance against Sweetpotato Feathery Mottle Virus." In *Scientist and Farmer: Partners in Research for the 21st Century*, CIP Program Report 1999–2000, 267–71. Lima: International Potato Centre, 2001.

Clapp, Jennifer. "The Political Economy of Food Aid in an Era of Agricultural Biotechnology." *Global Governance* 11, no. 4 (2005): 467–85. http://www.jstor.org/stable/27800586.

Cleaver, Harry. "The Contradictions of the Green Revolution." *American Economic Review* 62, no. 2 (1972): 177. ISSN: 0002-8282.

Cleveland, David A., and Daniella Soleri. "Rethinking the Risk Management Process for Genetically Engineered Crop Varieties in Small-Scale, Traditionally Based Agriculture." *Ecology and Society* 10, no. 1 (2005): 1–33. http://www.ecologyandsociety.org/vol10/iss1/art9.

Cohen, Joel and Robert Paarlberg. "Unlocking Crop Biotechnology in Developing Countries: A Report from the Field." *World Development* 32, no. 9 (2004): 1,563–77. http://doi.org/10.1016/j.worlddev. 2004.05.003.

Conrow, Joan. "Burkina Faso Puts GM Cotton on Hold." Cornell Alliance for Science, 27 April 2016. https://allianceforscience.cornell.edu/blog/2016/04/burkina-faso-puts-gm-cotton-on-hold.

– "Tanzania Plants Its First GMO Research Crop." *Cornwall Alliance for Science*, 5 October 2016. http://allianceforscience.cornell.edu/blog/tanzania-plants-its-first-gmo-research-crop.

Conway, Gordon. "From the Green Revolution to the Biotechnology Revolution: Food for Poor People in the 21st Century." Woodrow Wilson International Center for Scholars, director's forum, 12 March 2003.

Cook, Lynn J. "While the West Debates the Ethics of Genetically Modified Food, Florence Wambugu Is Using It to Feed Her Country." *Forbes*, 23 December 2002, https://www.forbes.com/free_forbes/2002/1223/302.html.

Cotton South Africa. "Textile Statistics." N.d. http://cottonsa.org.za/resources/textile-statistics.

– "Small-Scale Farmer Production Estimates," 2011. http://www.cottonsa.org.za.

Crane, Todd A. "Bringing Science and Technology Studies into Agricultural Anthropology: Technology Development as Cultural Encounter between Farmers and Researchers." *Culture, Agriculture, Food and Environment* 36, no. 1 (2014): 45–55. http://doi.org/10.1111/cuag.12028.

Croppenstedt, Andre, Markus Goldstein, and Nina Rosas. "Gender and Agriculture: Inefficiencies, Segregation, and Low Productivity Traps." *World Bank Research Observer* 28, no. 1 (2013): 79–109. http://doi.org/10.1093/wbro/lks024.

Cyranoski, David. "CRISPR Tweak May Help Gene-Edited Crops Bypass Biosafety Regulation." *Nature News*, 19 October 2015. http://www.nature.com/news/crispr-tweak-may-help-gene-edited-crops-bypass-biosafety-regulation-1.18590.

Dale, James, Paul Jean-Yves, Benjamin Dugdale, and Robert Harding. "Modifying Bananas: From Transgenics to Organics?" *Sustainability* 9, no. 3 (2017). http://doi.org/10.3390/su9030333.

Dalrymple, Dana. *Development and Spread of High-Yielding Varieties of Wheat and Rice in the Less Developed Nations.* Foreign Agricultural

Economic Report no. 95. Washington: United States Department of
Agriculture, 1978.

Dano, Elenita C. *Unmasking the New Green Revolution in Africa:
Motives, Players and Dynamics*. Third World Network, Church
Development Service and African Centre for Biosafety, 2007. http://
www.twn.my/title2/par/Unmasking.the.green.revolution.pdf.

deGrassi, Aaron. *Genetically Modified Crops and Sustainable Poverty
Alleviation in Sub-Saharan Africa: An Assessment of Current Evidence*.
Third World Network, 2003.

De Groote, Hugo. "Crop Biotechnology in Developing Countries." In
Plant Biotechnology and Agriculture, edited by Arie Altman and Paul
M. Hasegawa, 563–76. Oxford: Academic Press (Elsevier Press), 2011.

De Groote, Hugo, Getachew Dema, George B. Sonda, and Zachary M.
Gitonga. "Maize for Food and Feed in East Africa: The Farmers'
Perspective." *Field Crops Research* 153, C (2013): 22–36. http://doi.
org/10.1016/j.fcr.2013.04.005.

De Groote, Hugo, William Overholt, James Okuro Ouma, and Stephen
Mugo. "Assessing the Impact of Bt Maize in Kenya Using a GIS Model."
Paper presented at the 25th International Conference of the
International Agricultural Economics Association Conference, Durban,
South Africa, 16–22 August 2003.

Delve, Robert, Rui Benfica, Keizire B. Boaz, Joseph Rusike, Rebbie
Harawa, George Bigirwa, Fred Muhhuku, Jane Ininda, and John
Wakiumu. "Agricultural Productivity through Intensification and Local
Institutions." In *Africa Agriculture Status Report 2016: Progress
Towards Agricultural Transformation in Africa*, 108–127. Nairobi:
Alliance for a Green Revolution in Africa, 2016.

Demers-Morris, Cassandra. "GEOs and Gender: GEOs and What They
Mean for Women Farmers in Kenya." Master's dissertation, Dalhousie
University, 2015.

Department of Agriculture, Forest and Fisheries, Republic of South Africa.
"Maize Profile." N.d. http://www.nda.agric.za/docs/FactSheet/maize.htm.

Derera, John, and Tatenda Musimwa. "Why SR52 Is Such a Great Maize
Hybrid? I. Heterosis and Generation Mean Analysis." *Euphytica* 205,
no. 1 (2015): 121–35. http://doi.org/10.1007/s10681-015-1410-7.

De Schnutter, Olivier and Gaëtan Vanloqueren. "The New Green
Revolution: How Twenty-First-Century Science Can Feed the World."
Solutions for a Sustainable and Desirable Future 2, no. 4 (2011): 3.

Dessauw, Dominique, and Bernard Hau. "Cotton Breeding in French-
Speaking Africa: Milestones and Prospects." Paper presented at the

Omnipress World Cotton Research Conference 4, Lubbock, Texas, 2008.

Diallo, Alpha, and Kevin Pixley. CIMMYT in Sub-Saharan Africa: Weaving the Fabric Of Better Livelihoods. Harare, Zimbabwe: International Maize and Wheat Improvement Center (CIMMYT), 2003.

Diao, Xinshen, Derek Headey, and Michael Johnson. "Toward a Green Revolution in Africa: What Would It Achieve, and What Would It Require?" Agricultural Economics 39, no. 3 (2008): 539–50. http://doi.org/10.1111/j.1574-0862.2008.00358.x.

Dibden, Jacqui, David Gibbs, and Chris Cocklin. "Framing GM Crops as a Food Security Solution." Journal of Rural Studies, 29 (2013): 59–79. http://doi.org/10.1016/j.jrurstud.2011.11.001.

Doss, Leyla. "Environmentalists Protest against GM Seeds Multinational." Egypt Monocle, 28 May 2013. http://egyptmonocle.com/EMonocle/egypt-activists-march-against-monsanto.

Dowd-Uribe, Brian. "Engineered Outcomes: The State and Agricultural Reform in Burkina Faso." PhD dissertation, University of California, 2011.

– "Engineering Yields and Inequality? How Institutions and Agro-ecology Shape Bt Cotton Outcomes in Burkina Faso." Geoforum 53, C (2014): 161–71. http://doi.org/10.1016/j.geoforum.2013.02.010.

East African. "Tanzania Plans April Trials for GMO Maize Varieties," 27 October 2015. http://allafrica.com/stories/201510271523.html.

EcoBank. "Burkina Faso: Changes Loom for 2015/16 Cotton Season." Middle Africa Briefing Note, Soft Commodities, Cotton, 12 June 2015. http://www.ecobank.com/upload/201506150552098129863cYnb2gu98T.pdf.

El-Shamei, Zakarya S., Amal A. Gab-Alla, Adel Shatta, Eid Moussa, and Ahmed Mohamed Rayan. "Histopathological Changes in some Organs of Male Rats Fed on Genetically Modified Corn (Ajeeb YG)." Journal of American Science 8, no. 10 (2012): 684–96. ISSN: 1545-1003.

El Amin, El Tigani M., and Musa A. Ahmed. "Strategies for Integrated Pest Control in the Sudan: 1. Cultural and Legislative Measures." International Journal of Tropical Insect Science 12 (1991): 547–52. ISSN: 0191-9040.

El Wakeel, Ahmed S. "Agricultural Biotechnology and Biosafety Regulations in Sudan: The Case of Bt Cotton." Paper presented at the NASAC Agricultural Biotechnology Workshop Promoting Agricultural Biotechnology for Sustainable Development in Africa, Addis Ababa, 25–6 February 2014.

Elbehri, Aziz, and Steve Macdonald. "Estimating the Impact of Transgenic Bt Cotton on West and Central Africa: A General Equilibrium Approach." *World Development* 32, no. 12 (2004): 2,049–64. http://doi.org/10.1016/j.worlddev.2004.07.005.

Erasmus, Annemie, Jaco Marais, and Johnnie Van Den Berg. "Movement and Survival of *Busgfseola fusca* (Lepidoptera: Noctuidae) Larvae within Maize Plantings with Different Ratios of Non-Bt and Bt Seed." *Pest Management Science* 72, no. 12 (2016): 2287–94. http://doi.org/10.1002/ps.4273.

Eric Holt-Giménez. "Out of AGRA: The Green Revolution Returns to Africa." *Development* 51, no. 4 (2008): 464–71. http://doi.org/10.1057/dev.2008.49.

Estur, Gérald. "Quality and Marketing of Cotton Lint in Africa." Africa Region working paper series no. 121, World Bank, Washington, DC, 2008.

European Food Safety Authority. "GMO." N.d. https://www.efsa.europa.eu/en/science/gmo.

Eveleens, Kees. "Cotton-Insect Control in the Sudan Gezira: Analysis of a Crisis." *Crop Protection* 2, no. 3 (1983): 273–87. http://doi.org/10.1016/0261-2194(83)90002-9.

Evenson, Robert E., and Douglas Gollin. "Assessing the Impact of the Green Revolution, 1960 to 2000." *Science* 300, no. 5,620 (2003): 758–62. ISSN:0036-8075.

Ezezika, Obidimma C., and Abdallah S. Daar. "Building Trust in Biotechnology Crops in Light of the Arab Spring: A Case Study of Bt Maize in Egypt." *Agriculture and Food Security* 1, no. 1 (2012): 1–7. ISSN: 2048-7010.

Ezezika, Obidimma C., Jennifer Deadman, and Abdallah S. Daar. "She Came, She Saw, She Sowed: Re-negotiating Gender-Responsive Priorities for Effective Development of Agricultural Biotechnology in Sub-Saharan Africa." *Journal of Agricultural and Environmental Ethics* 26, no. 2 (2013): 461–71. http://doi.org/10.1007/s10806-012-9396-9.

Faber, Mieke, and Paul J. van Jaarsveld. "The Production of Provitamin A-Rich Vegetables in Home-Gardens as a Means of Addressing Vitamin A Deficiency in Rural African Communities." *Journal of the Science of Food and Agriculture* 87, no. 3 (2007): 366–77. http://doi.org/10.1002/jsfa.2774.

Faber, Mieke, and Sunette M. Laurie. "A Home Gardening Approach Developed in South Africa to Address Vitamin A Deficiency." In *Combating Micronutrient Deficiencies: Food-Based Approaches*, edited by Brian Thompson and Leslie Amoroso, 163–84. Rome: Food and Agriculture Organization of the United Nations and CABI, 2011.

Fagan, John, Terje Traavik, and Thomas Bøhn. "The Seralini Affair: Degeneration of Science to Re-Science?" *Environmental Sciences Europe* 27, no. 1 (2015): 1–9. http://doi.org/10.1186/s12302-015-0049-2.

Falck-Zapeda, Jose, and Patricia Zambrano. *Gender Impacts of Genetically Engineered Crops in Developing Countries: Final Technical Paper.* Washington, DC: International Development Research Centre, 2013.

FAO (Food and Agricultural Organisation of the United Nations). "Statistics Division." N.d. http://faostat3.fao.org/download/Q/QC/E.

– "FAOSTAT." N.d. http://www.fao.org/faostat.

– *The State of Food and Agriculture 2003–2004: Agricultural Biotechnology Meeting the Needs of the Poor?* Rome: Food and Agriculture Organization, 2004.

– *Smallholders and Family Farmers.* Rome: Food and Agriculture Organization, 2012. http://www.fao.org/family-farming/detail/en/c/273864.

– *Sustainability Assessment of Food and Agricultural Systems Guidelines.* Rome: Food and Agriculture Organization, 2014.

Feed the Future. *The US Government's Global Food Security Research Strategy: Reducing Global Hunger, Malnutrition and Poverty Through Science, Technology and Innovation.* Feed the Future, 2017.

Ferguson, James. *The Anti-Politics Machine: "Development," Depoliticization, and Bureaucratic Power in Lesotho.* Cambridge, NY: Cambridge University Press, 1990.

Feris, Loretta. "Risk Management and Liability for Environmental Harm Caused by GMOs: The South African Regulatory Framework." *Potchefstroom Electronic Law Journal* 1 (2006): 1–26. ISSN: 17273781.

Fischer, Klara, and Flora Hajdu. "Does Raising Maize Yields Lead to Poverty Reduction? A Case Study of the Massive Food Production Programme in South Africa." *Land Use Policy* 46 (2015): 304–13. http://doi.org/10.1016/j.landusepol.2015.03.015.

Fischer, Klara, Elisabeth Ekener-Petersen, Lotta Rydhmer, and Karin Björnberg. "Social Impacts of GM Crops in Agriculture: A Systematic Literature Review." *Sustainability* 7, no. 7 (2015): 8598–620. http://doi.org/10.3390/su7078598.

Fischer, Klara, Johnnie Van Den Berg, and Charles Mutengwa. "Is Bt Maize Effective in Improving South African Smallholder Agriculture?"

South African Journal of Science 111, no. 1 (2015): 15–16. http://doi.org/10.17159/sajs.2015/a0090.

Flachs, Andrew. "Redefining Success: The Political Ecology of Genetically Modified and Organic Cotton as Solutions to Agrarian Crisis." *Journal of Political Ecology* 23, no. 1 (2016): 49–70. http://doi.org/10.2458/v23i1.20179.

Fleming, Nick. "Science's Search for a Super Banana." *Guardian*, 5 August 2018.

Fok, Michel, Marnus Gouse, Jean-Luc Hofs, and Johann Kirsten. "Contextual Appraisal of GM Cotton Diffusion in South Africa." *Life Science International Journal* 1, no. 4 (2007): 468–82.

Forsyth, Tim. *Critical Political Ecology: The Politics of Environmental Science*. New York: Routledge, 2003.

Foti, Richard, Celtos Mapiye, Mutenje Munyaradzi, Marizvikuru Mwale, and Nyararai Mlambo. "Farmer Participatory Screening of Maize Seed Varieties for Suitability in Risk Prone, Resource-Constrained Smallholder Farmer Systems of Zimbabwe." *African Journal of Agricultural Research* 3, no. 3 (2008): 180–5. ISSN 1991-637X.

Friedmann, Harriet. "Feeding the Empire: The Pathologies of Globalized Agriculture." *The Socialist Register 2005*, edited by Leo Panitch and Colin Leys, 124–43. London: Merlin Press, 2005.

Friedmann, Harriet, and Philip McMichael. "Agriculture and the State System: The Rise and Decline of National Agricultures, 1870 to the Present." *Sociologia Ruralis* 29, no. 2 (1989): 93–117. http://doi.org/10.1111/j.1467-9523.1989.tb00360.x

Fukuda-Parr, Sakiko. *The Gene Revolution: GM Crops and Unequal Development*. Earthscan London, 2006.

Gab-Alla, Amal A., Zakarya A. El-Shamei, Adel A. Shatta, Eid A. Moussa, and Ahmed M. Rayan. "Morphological and Biochemical Changes in Male Rates Fed on Genetically Modified Corn (Ajeeb YG)." *Journal of American Science* 8, no. 9 (2012): 117–23. ISSN: 1545-1003.

Gachenga, Elizabeth. "Status of Biotechnology and Biosafety Policies and Laws in COMESA Countries." Report presented at the Regional Workshop on Development of a Biotech/Biosafety Communication Strategy for the COMESA Region, 2008.

Galt, Ryan E. "Placing Food Systems in First World Political Ecology: A Review and Research Agenda." *Geography Compass*, 7 (2013): 637–58. http://doi.org/10.1111/gec3.12070.

Gao, Caixia, and Klaus K. Nielsen. "Comparison Between *Agrobacterium*-Mediated and Direct Gene Transfer using the Gene Gun." *Methods in*

Molecular Biology (Clifton, NJ) 940 (2013): 3–16. http://doi.org/
10.1007/978-1-62703-110-3_1.

Gareth, Austin. "Resources, Techniques, and Strategies South of the Sahara:
Revising the Factor Endowments Perspective on African Economic
Development, 1500–2000." *Economic History Review* 61, no. 3 (2008):
587–624. http://doi.org/10.1111/j.1468-0289.2007.00409.x.

Gasura, Edmore, Arnold Bray Mashingaidze, and Settumba B. Mukasa.
"Occurrence, Prevalence and Implications of Sweet Potato Recovery
from Sweet Potato Virus Disease in Uganda." *African Crop Science
Conference Proceedings* 9 (2009): 601–8. http://doi.org/10.1080/07060
661.2015.1004111.

Gathura, Gatonye. "GM Technology Fails Local Potatoes." *Daily Nation*,
29 January 2004.

Gengenbach, Heidi, Thomas Bassett, William Moseley, William Munro,
and Rachel Schurman. "Limits of the Green Revolution for Africa
(GR4A): Reconceptualizing Gendered Agricultural Value Chains."
Geographical Journal 183, no. 3 (2017): 208–14. http://doi.org/
10.1111/geoj.12233.

Gerson, Micheal. "Are You Anti-GMO? Then You're Anti-Science, Too."
Washington Post, 3 May 2018.

Ghana News Agency. "Cotton Development Authority Board Inaugurated."
3 April 2015. http://citifmonline.com/2015/04/03/cotton-development-
authority-board-inaugurated.

Gibson, Rosalind S., and Christine Hotz. "Dietary Diversification/
Modification Strategies to Enhance Micronutrient Content and
Bioavailability of Diets in Developing Countries." *British Journal of
Nutrition* 85, S2 (2001): S159–66. http://doi.org/10.1079/BJN2001309.

Giller, Ken, Jens A. Andersson, Marc Corbeels, John Kirkegaard, David
Mortensen, Olaf Erenstein, and Bernard Vanlauwe. "Beyond
Conservation Agriculture." *Frontiers in Plant Science* 6 (2015): 870.
http://doi.org/10.3389/fpls.2015.00870.

Giller, Ken, Jens A. Andersson, James Sumberg, and John Thompson. "A
Golden Age for Agronomy?" In *Agronomy for Development: The
Politics of Knowledge in Agricultural Research*, edited by James
Sumberg, 150–60. New York: Routledge, 2017.

Global Harvest Initiative. *2012 Gap Report: Measuring Global Agricultural
Productivity*. Washington, DC: Global Harvest Initiative, 2012.

Glover, Dominic. "Exploring the Resilience of Bt Cotton's 'Pro-Poor
Success Story.'" *Development and Change* 41, no. 6 (2010): 955–81.
http://doi.org/10.1111/j.1467-7660.2010.01667.x.

– "The System of Rice Intensification: Time for an Empirical Turn."
    NJAS-Wageningen Journal of Life Sciences 57, no. 3–4 (2011): 217–24.
    http://doi.org/10.1016/j.njas.2010.11.006.
Glover, Dominic, James Sumberg, and Jens A. Andersson. "The Adoption
    Problem; or Why We Still Understand So Little about Technological
    Change in African Agriculture." *Outlook on Agriculture* 45, no. 1
    (2016): 3–6. http://doi.org/10.5367/oa.2016.0235.
Glover, Dominic, Jean-Philippe Venot, and Harro Maat. "On the
    Movement of Agricultural Technologies: Packaging, Unpacking and
    Situated Reconfiguration." In *Agronomy for Development: The Politics
    of Knowledge in Agricultural Research*, edited by James Sumberg,
    14–30. New York: Routledge, 2017.
GM Watch. "Burkina Faso Abandons GM Bt Cotton," 2016. https://www.
    gmwatch.org/en/2016-articles/16677-burkina-faso-abandons-gm-
    bt-cotton.
Gogo, Jeffrey. "Zimbabwe: Food Shortages Re-Ignite GMO Concerns."
    The Herald, 11 April 2016. http://allafrica.com/stories/201604110243.
    html.
Gouse, Marnus. "GM Maize as Subsistence Crop: The South African
    Smallholder Experience." *AgBioForum* 15, no. 2 (2012): 163–74.
– *Seed Technology and Production System Comparisons: South African
    Subsistence/Smallholder Farmers*, 2014. http://www.community
    commons.org/wp-content/uploads/bp-attachments/37350/Gouse-seed-
    tech-comp-report-final.pdf.
Gouse, Marnus, Johann F. Kirsten, and Lindie Jenkins. "Bt Cotton in
    South Africa: Adoption and the Impact of Farm Incomes Amongst
    Small-Scale and Large-Scale Farmers." *Agrekon* 42, no. 1 (2003):
    15–28. http://doi.org/10.1080/03031853.2003.9523607.
Gouse, Marnus, Johann F. Kirsten, and Wynand J. van der Walt. *Bt Cotton
    and Bt Maize: An Evaluation of Direct and Indirect Impact on the
    Cotton and Maize Farming Sectors in South Africa*. Pretoria, South
    Africa, 2008.
Gouse, Marnus, Jenifer Piesse, Colin Thirtle, and Colin Poulton.
    "Assessing the Performance of GM Maize Amongst Smallholders in
    KwaZulu-Natal, South Africa." *AgBioForum* 12 (2009): 78–89. ISSN:
    1522-936X.
Gouse, Marnus, Carl Pray, Johann Kirsten, and David Schimmelpfennig.
    "A GM Subsistence Crop in Africa: The Case of Bt White Maize in
    South Africa." *International Journal of Biotechnology* 7 (2005): 84–94.
    http://doi.org/10.3390/su8090865.

- "Three Seasons of Subsistence Insect-Resistant Maize in South Africa: Have Smallholders Benefited?" *AgBioForum* 9, no. 1 (2006): 15–22.

Gouse, Marnus, Debdatta Sengupta, Patricia Zambrano, and Jose Falck Zepeda. "Genetically Modified Maize: Less Drudgery for Her, More Maize for Him? Evidence from Smallholder Maize Farmers in South Africa." *World Development* 83 (2016): 27–38. http://doi.org/10.1016/j.worlddev.2016.03.008.

Gouse, Marnus, Bhavani Shankar, and Colin Thirtle. "The Decline of Bt Cotton in KwaZulu-Natal: The Development and Spread of GM Cotton." In *Hanging By a Thread: Cotton, Globalization, and Poverty in Africa*, edited by William G. Moseley and Leslie Gray, 103–20. Athens, OH: Ohio University Press, 2008.

GRAIN. *USAID: Making the World Hungry for GM Crops*, 2005. https://www.grain.org/article/entries/21-usaid-making-the-world-hungry-for-gm-crops.

Gray, Leslie, and Brian Dowd-Uribe. "A Political Ecology of Socio-Economic Differentiation: Debt, Inputs and Liberalization Reforms in Southwestern Burkina Faso." *Journal of Peasant Studies* 40, no. 4 (2013): 683–702. http://doi.org/10.1080/03066150.2013.824425.

Green, Erik. "Production Systems in Pre-colonial Africa." In *The History of African Development: An Online Textbook for a New Generation of African Students and Teachers*, edited by Ewout Frankema, Ellen Hillbom, Ushehwedu Kufakurinani, and Felix Meier zu Selhausen. African Economic History Network, 2013.

Groote, Hugo, William Overholt, James Ouma, and Japhether Wanyama. "Assessing the Potential Economic Impact of *Bacillus thuringiensis* (Bt) Maize in Kenya." *African Journal of Biotechnology* 10, no. 23 (2011): 4,741-51. ISSN: 1684-5315.

Grow Africa. *Grow Africa: Partnering for Agricultural Transformation*. Johannesburg, South Africa: Grow Africa, 2017. https://www.growafrica.com/sites/default/files/20170113-Grow%20Africa-Corporate%20Flyer-WEB_2.pdf.

Gurian-Sherman, Doug. *High and Dry: Why Genetic Engineering Is Not Solving Agriculture's Drought Problem in a Thirsty World*. Cambridge, MA: Union of Concerned Scientists, 2012.

Haakonsson, Stine, Johanna Gammelgaard, and Sine N. Just. "Corporate Scramble for Africa?: Exploring the Interrelations of International Business and Politics in the Case of New Alliance for Food Security and Nutrition." Paper presented at the 33rd EGOS Colloquium, 7 July 2017, Copenhagen.

Hahn, S.K., and Janet Keyser. "Cassava: A Basic Food of Africa." *Outlook on Agriculture* 14, no. 2 (1985): 95–100. http://doi.org/10.1177/003072708501400207.

Hanna, Simon. "Video: Egypt Activists Protest Monsanto's GM Seeds." *Ahram Online*, 15 May 2013. http://english.ahram.org.eg/NewsContentMulti/72306/Multimedia.aspx.

Harsh, Matthew, and James Smith. "Technology, Governance and Place: Situating Biotechnology in Kenya." *Science and Public Policy* 34, no. 4 (2007): 251–60. http://doi.org/10.3152/030234207X214444.

Hazell, Peter, Colin Poulton, Steve Wiggins, and Andrew Dorward. *The Future of Small Farms for Poverty Reduction and Growth.* 2020 discussion paper no. 42. Washington, DC: International Food Policy Research Institute, 2007.

Herring, Ronald J. "The Genomics Revolution and Development Studies: Science, Poverty and Politics." *Journal of Development Studies* 43, no. 1 (2007): 1–30. http://doi.org/10.1080/00220380601055502.

High Level Panel of Experts (HLPE). *Investing in Smallholder Agriculture for Food Security.* High Level Panel of Experts on Food Security and Nutrition of the Committee on World Food Security, Rome, 2013.

Hillocks, Rory J. "Is There a Role for Bt Cotton in IPM for Smallholders in Africa?" *International Journal of Pest Management* 51, no. 2 (2005): 131–41. http://doi.org/10.1080/09670870500117292.

Hoag, Hannah. "Biotech Firms Join Charities in Drive to Help Africa's Farms." *Nature* 422, no. 6,929 (2003): 246. http://doi.org/10.1038/422246a.

Hockin-Grant, Kenneth, and Yasué, Maï. "The Effectiveness of a Permaculture Education Project in Butula, Kenya." *International Journal of Agricultural Sustainability* 15, no. 4 (2017): 432–44. http://doi.org/10.1080/14735903.2017.1335570.

Hodge, Joseph Morgan. *Triumph of the Expert: Agrarian Doctrines of Development and the Legacies of British Colonialism.* Athens, OH: Ohio University Press, 2007.

Hofs, Jean-Luc, Michel Fok, Marnus Gouse, and Johann Kirsten. "Diffusion du coton génétiquement modifié en Afrique du Sud: des leçons pour l'Afrique Zone Franc." *Revue Tiers Monde,* 188 (2006): 773–6. http://doi.org/10.3406/tiers.2006.6463.

Horna, Daniela, Miriam Kyotalimye, and Jose Falck-Zepeda. "Cotton Production in Uganda: Would GM Technologies Be the Solution?" Paper presented at the International Association of Agricultural Economists Conference, Beijing, China, 16–22 August 2009. http://ageconsearch.

umn.edu/bitstream/51823/2/Horna-Kyotalymye-Falck-Zepeda-JUN
30th.pdf.

Horna, Daniela, Patricia Zambrano, and Jose Falck-Zepeda. *Socio-economic Considerations in Biosafety Decision Making: Methods and Implementation.* Washington, DC: International Food Policy Research, 2013.

Horsch, Robert. "Plant Biotechnology Research in Africa." Congressional testimony by Robert Horsch, Monsanto, 12 June 2003. http://www.agbioworld.org/newsletter_wm/index.php?caseid=archive&newsid=1702.

Horsch, Robert, and Jill Montgomery. "Why We Partner: Collaborations Between the Private and Public Sectors for Food Security and Poverty Alleviation through Agricultural Biotechnology." *AgBioForum* 7, no. 1 (1994): 80–3. ISSN: 1522-936X.

Hoyer, Ute, Edgar Maiss, Wilhelm Jelkmann, and Dietrich E. Lesemann. "Identification of the Coat Protein Gene of a Sweet Potato Sunken Vein Closterovirus Isolate from Kenya and Evidence for a Serological Relationship among Geographically Diverse Closterovirus Isolates from Sweet Potato." *Phytopathology* 86, no. 7 (1996): 744–50. http://doi.org/10.1094/Phyto-86-744.

Huang, Fangneng, Jawwad A. Qureshi, Robert L. Meagher Jr, Dominic D. Reisig, Graham P. Head, David A. Andow, Xinzi Ni, David Kerns, G. David Buntin, Ying Niu, Fei Yang, and Vikash Dangal. "Cry1F Resistance in Fall Armyworm *Spodoptera frugiperda*: Single Gene Versus Pyramided Bt Maize." *PLoS One* 9, no. 11 (2014): e112958. http://doi.org/10.1371/journal.pone.0112958.

Hussein, Marwa. "Egyptian Activists Launch First Protest against Genetically Modified Food." *Ahram Online*, 26 May 2013. http://english.ahram.org.eg/News/72305.aspx.

IFAD (International Fund for Agricultural Development). *Rural Development Report 2016: Fostering Inclusive Rural Transformation.* Rome: International Fund for Agricultural Development, 2016.

ISAAA (International Service for the Acquisition of Agri-biotech Applications). "ISAAA AfriCenter." N.d. http://africenter.isaaa.org.

– "ISAAA in 2016." N.d. http://www.isaaa.org/resources/publications/annualreport/2016/default.asp.

– "Sudanese Farmers Reap Benefits from Bt Cotton." *Crop Biotech Update*, 19 November 2014. http://www.isaaa.org/kc/cropbiotech update/article/default.asp?ID=12904.

– *Agricultural Biotechnology (A Lot More than Just GM Crops).* Laguna, Philippines: ISAAA SEAsiaCenter, 2014.

– "Tanzania Deputy Minister for Agri Advocates Adoption of Biotech in
  Agriculture." *Crop Biotech Update*, 1 July 2015. http://www.isaaa.org/
  kc/cropbiotechupdate/article/default.asp?ID=13522.
Isaacman, Allen F., and Richard L. Roberts. *Cotton, Colonialism, and
  Social History in Sub-Saharan Africa*. Portsmouth, NH: Heinemann,
  1995.
Ismael, Yousouf, Richard Bennett, and Stephen Morse. "Benefits from Bt
  Cotton Use by Smallholder Farmers in South Africa." *AgBioForum* 5,
  no. 1 (2002): 1–5.
Iversen, Marianne, Idun Grønsberg, Johnnie Berg, Klara Fischer, Denis W.
  Aheto, and Thomas Bøhn. "Detection of Transgenes in Local Maize
  Varieties of Small-Scale Farmers in Eastern Cape, South Africa." *PLoS
  ONE* 9, no. 12 (2014): 1–21. http://doi.org/10.1371/journal.
  pone.0116147.
Jacobson, Klara. *From Betterment to Bt: Agricultural Development and
  the Introduction of Genetically Modified Maize to South African
  Smallholders*. Uppsala: Swedish University of Agricultural Sciences,
  2013. http://pub.epsilon.slu.se/10406.
Jacobson, Klara, and Anne Ingeborg Myhr. "GM Crops and Smallholders."
  Journal of Environment and Development 22, no. 1 (2013): 104–24.
  http://doi.org/10.1177/1070496512466856.
Jaffe, Gregory. "Ensuring Biosafety at the National Level: Suggestions to
  Improve the Operation of the South African Biosafety Regulatory
  System." Program for Biosafety Systems (PBS), International Food
  Policy Research Institute (IFPRI), PBS brief number 11, 2008. https://
  www.cbd.int/doc/external/mop-04/ifpri-pbs-policy-11-en.pdf.
Jain, Shri Mohan. "Mutation Induced Genetic Improvement." In *New
  Approaches to Plant Breeding of Orphan Crops in Africa*, edited by
  Zerihun Tadele, 115–26. Bern, Switzerland, 2007.
– "Mutation Induced Genetic Improvement." In *New Approaches to Plant
  Breeding of Orphan Crops in Africa: Proceedings of an International
  Conference*, edited by Zerihun Tadele, 115–26. Bern, Switzerland, 2009.
Jarosz, Lucy. "Growing Inequality: Agricultural Revolutions and the
  Political Ecology of Rural Development." *International Journal of
  Agricultural Sustainability* 10, no. 2 (2011): 192–9. http://doi.org/10.10
  80/14735903.2011.600605.
Jogo, Wellington, Eldad Karamura, William Tinzaara, Jerome Kubiriba,
  and Anne Rietveld. "Determinants of Farm-Level Adoption of Cultural
  Practices for Banana Xanthomonas Wilt Control in Uganda." *Journal of
  Agricultural Science* 5, no. 7 (2013): 70–81. http://doi.org/10.5539/jas.
  v5n7p70.

Jozini Municipality. *Jozini Local Municipality* IDP *Review (2005/2006)*. First draft.

Juma, Calestous, and Ismail Serageldin. 2007. *Freedom to Innovate: Biotechnology in Africa's Development – Report of the High-Level African Panel on Modern Biotechnology. High-Level African Panel on Modern Biotechnology*. Addis Ababa and Pretoria: African Union (AU) and New Partnership for Africa's Development (NEPAD), 2004.

Just, David R., and Harry M. Kaiser. "A Good Deal for the Environment and the Poor." *Forbes*, 14 September 2016.

Kabunga, Nassul, Thomas Dubois, and Matin Qaim. "Impact of Tissue Culture Banana Technology on Farm Household Income and Food Security in Kenya." *Food Policy* 45 (2014): 25–34. http://doi.org/10.1016/j.foodpol.2013.12.009.

Kalaitzandonakes, Nicholas, John Kruse, and Marnus Gouse. "The Potential Economic Impacts of Herbicide-Tolerant Maize in Developing Countries: A Case Study." *AgBioForum* 18, no. 2 (2015): 221–38.

Kalule, T., Z. Khan, G. Bigirwa, J. Alupo, S. Okanya, J. Pickett, and L. Wadhams. "Farmers' Perceptions of Importance, Control Practices and Alternative Hosts of Maize Stemborers in Uganda." *International Journal of Tropical Insect Science* 26, no. 2 (2006): 71–7. http://doi.org/10.1079/IJT2006103.

Kandiwa, Vongai, and Amare Tegbaru. "Gender in DTMA." In *DTMA Highlights for 2012/13*, edited by Tsedeke Abate, Abebe Menkir, John F. MacRobert, Girma Tesfahun, Tahirou Abdoulaye, Peter Setimela, Baffour Badu-Apraku, Dan Makumbi, Cosmos Magorokosho, Amsal Tarekegne, 68–71. Kenya: CIMMYT, 2013.

Kangmennaang, Joseph, Rachel Bezner Kerr, Esther Lupafya, Laifolo Dakishoni, Mangani Katundu, and Isaac Luginaah. "Impact of a Participatory Agroecological Development Project on Household Wealth and Food Security in Malawi." *Food Security* 9, no. 3 (2017): 561–76. http://doi.org/10.1007/s12571-017-0669-z.

Karamura, Eldad, and William Tinzaara. "Management of Banana Xanthomonas Wilt in East and Central Africa." Proceedings of the Workshop on Review of the Strategy for the Management of Banana Xanthomonas Wilt, 23–7 July 2007, Hotel la Palisse, Kigali, Rwanda. http://banananetworks.org/barnesa/files/2013/05/Untitled1.pdf.

Karamura, Eldad, G. Kayobyo, W. Tushemereirwe, S. Benin, Guy Blomme, S. Eden Green, and R. Markham. "Assessing the Impacts of Banana Bacterial Wilt Disease on Banana (Musa Spp.) Productivity and Livelihoods of Ugandan Farm Households." *Acta Horticulturae* 8,792 (2010): 749–55. http://doi.org/10.17660/ActaHortic.2010.879.81.

Karembu, Margaret, Faith Nguthi, and Ismail Abdel-Hamid. *Biotech Crops in Africa: The Final Frontier*. Nairobi, Kenya: ISAAA, 2009.

KARI/CIMMYT. *Insect Resistant Maize for Africa (IRMA) Project Annual Report 2003–2004*. IRMA project document no. 20, 2004.

Karyeija, Robert F., Richard W. Gibson, and Jari P.T. Valkonen. "Resistance to Sweet Potato Virus Disease (SPVD) in Wild East African Ipomoea." *Annals of Applied Biology* 133, no. 1 (1998): 39–44. http://doi.org/10.1111/j.1744-7348.1998.tb05800.x.

Karyeija, Robert F., Jan F. Kreuze, Richard W. Gibson, and Jari P.T. Valkonen. "Synergistic Interactions of a Potyvirus and a Phloem-Limited Crinivirus in Sweet Potato Plants." *Virology* 269, no. 1 (2000): 26–36. http://doi.org/10.1006/viro.1999.0169.

– "Variability of Sweetpotato Feathery Mottle Virus in Africa." *African Crop Science Journal* 9, no. 1 (2001): 293–9. http://doi.org/10.4314/acsj.v9i1.27651.

Keeley, James, and Ian Scones. *Contexts for Regulations: GMOs in Zimbabwe*. Brighton: Institute of Development Studies, University of Sussex, 2003.

Keetch, David, Diran Makinde, Cholani K. Weebadde, and Karim M. Maredia. *Biosafety in Africa: Experiences and Best Practices*. East Lansing: Michigan State University, 2014.

Keetch, David, J. Webster, A. Ngqaka, R. Akanbi, and P. Mahlanga. "Bt Maize for Small Scale Farmers: A Case Study." *African Journal of Biotechnology* 4, no. 13 (2005): 1,505–9. ISSN 1684-5315.

Kent, Laurence. "Bringing Drought Tolerance to Africa: The Bill and Melinda Gates Foundation's Support of Biotechnology." *RealAgriculture*, 8 October 2014. Transcribed from video file. https://www.realagriculture.com/2014/10/bill-melinda-gates-foundations-role-ag-biotech.

Kharkwal, M.C., and Q.Y. Shu. "The Role of Induced Mutations in World Food Security." In *Induced Plant Mutations in the Genomics Era*, edited by Q. Y. Shu, 33–38. Rome: Food and Agriculture Organization of the United Nations (FAO) and the International Atomic Energy Agency (IAEA), 2009. http://www.fao.org/docrep/012/i0956e/I0956e.pdf

Khatodia, Surender, Kirti Bhatotia, Nishat Passricha, S.M.P. Khurana, and Narendra Tuteja. "The CRISPR/Cas Genome-Editing Tool: Application in Improvement of Crops." *Frontiers in Plant Science* 7 (2016): 506. http://doi.org/10.3389/fpls.2016.00506.

Khush, Gurdev. "Rice Breeding: Past, Present and Future." *Journal of Genetics* 66, no. 3 (1987): 195–216. http://doi.org/10.1007/BF02927713.

Khush, Gurdev. "Rice Breeding: Past, Present and Future." *Journal of Genetics* 66, no. 3 (1987): 195–216. http://doi.org/10.1007/BF02927713.

— "Challenges for Meeting the Global Food and Nutrient Needs in the New Millennium." *Proceedings of the Nutrition Society* 60, no. 1 (2001): 15–26. http://doi.org/10.1079/PNS200075.

Kikafunda, Joyce Kakuramatsi, Estelle Bader, Giulia Palma, Maylis Razès, and Marie Claude Dop. "Nutrition Country Profile the Republic of Uganda." Rome: Food and Agricultural Organization of the United Nations (FAO), 2010. http://www.fao.org/3/a-bc643e.pdf.

Kikulwe, Enoch. "Banana Tissue Culture: Community Nurseries for African Farmers." In *Case Studies of Roots, Tubers and Bananas Seed Systems*, 180–96 RTB working paper 2016-3, CGIAR Research Program on Roots, Tubers and Bananas, Lima, Peru, 2016.

Kireger, Eliud. Press conference comments. Kenya Agricultural and Livestock Research Organization, n.d. http://wema.AATF-africa.org/files/files/project_sites_publications/KALRO-DG_press-conference-comments.pdf.

Kloppenburg, Jack Ralph. *First the Seed: The Political Economy of Plant Biotechnology, 1492–2000*. 2nd ed. Madison, WI: University of Wisconsin Press, 2004.

Kostandini, Genti, Bradford F. Mills, and Elton Mykerezi. "*Ex-Ante* Evaluation of Drought Tolerant Varieties in Eastern and Central Africa." *Journal of Agricultural Economics* 62, no. 1 (2011): 172–206. http://doi.org/10.1111/j.1477-9552.2010.00281.x.

Kostandini, Genti, Roberto La Rovere, and Tahirou Abdoulaye. "Potential Impacts of Increasing Average Yields and Reducing Maize Yield Variability in Africa." *Food Policy* 43 (2013): 213–22. http://doi.org/10.1016/j.foodpol.2013.09.007.

Kruger, Marlene, J.B.J. Van Rensburg, and Johnnie Van Den Berg. "Perspective on the Development of Stem Borer Resistance to Bt Maize and Refuge Compliance at the Vaalharts Irrigation Scheme in South Africa." *Crop Protection* 28, no. 8 (2009): 684–9. http://doi.org/10.1016/j.cropro.2009.04.001.

— "Resistance to Bt Maize in *Busseola fusca* (Lepidoptera: Noctuidae) from Vaalharts, South Africa." *Environmental Entomology* 40, no. 2 (2011): 477–83. http://doi.org/10.1603/EN09220.

Kubiriba, Jerome, Erima Rockefeller, and Wilberforce K. Tushemereirwe. "Scaling Out Control of Banana Xanthomonas Wilt from Community to Regional Level: A Case from Uganda's Largest Banana Growing Region." *Journal of Development and Agricultural Economics* 8, no. 5 (2016): 108–17. http://doi.org/10.5897/JDAE2014.0623.

Lacombe, Camille, Nathalie Couix, and Laurent Hazard. "Designing
    Agroecological Farming Systems with Farmers: A Review." *Agricultural
    Systems* 165 (2018): 208–20. http://doi.org/10.1016/j.agsy.2018.06.014.
Lau, Phing, Wendy Chui, Mohammad Abdul Latif, Mohd Y. Rafii, Mohd
    R. Ismail, and Adam Puteh. "Advances to Improve the Eating and
    Cooking Qualities of Rice by Marker-Assisted Breeding." *Critical
    Reviews in Biotechnology* 36, no. 1 (2016): 87–98. http://doi.org/10.31
    09/07388551.2014.923987.
Le Chalony, Catherine, and Jean-Yves Moisseron. "Research Governance
    in Egypt: Biotechnology as a Case Study." *Science, Technology and
    Society* 15, no. 2 (2010): 371–97. http://doi.org/10.1177/09717218100
    1500208.
Lemgo, Godwin, and Samuel Timpo. "Case Study: Ghana." In *Biosafety in
    Africa: Experiences and Best Practices*, edited by David P. Keetch, Diran
    Makinde, Cholani K. Weebadde, and Karim M. Maredia, 100–9. East
    Lansing, MI: Michigan State University, 2014. http://nepad-abne.net/
    wp-content/uploads/2015/11/Biosafety-Africa-PartIv2.pdf.
Lescourret, Françoise, Thierry Dutoit, Freddy Rey, François Côte,
    Marjolaine Hamelin, and Eric Lichtfouse. "Agroecological Engineering."
    Agronomy for Sustainable Development 35, no. 4 (2015): 1,191–8.
Li, Ting, Bo Liu, Martin H. Spalding, Donald P. Weeks, and Yang Bing.
    "High-Efficiency TALEN-Based Gene Editing Produces Disease-
    Resistant Rice." *Nature Biotechnology* 30, no. 5 (2012): 390–2. http://
    doi.org/10.1038/nbt.2199.
Licker, Rachel, Matt Johnston, Jonathan A. Foley, Carol Barford,
    Christopher J. Kucharik, Chad Monfreda, and Navin Ramankutty.
    "Mind the Gap: How Do Climate and Agricultural Management
    Explain the 'Yield Gap' of Croplands around the World?" *Global
    Ecology and Biogeography* 19, no. 6 (2010): 769–82. http://doi.org/
    10.1111/j.1466-8238.2010.00563.x.
Lieberman, Sarah, and Tim Gray. "GMOs and the Developing World: A
    Precautionary Interpretation of Biotechnology." *British Journal of
    Politics and International Development* 10, no. 3 (2008): 395–411.
    http://doi.org/10.1111/j.1467-856X.2007.00304.x.
Ligami, Christabel. "Bacterial Wilt-Resistant Banana Variety in the
    Offing." *East African Observer*, 15 June 2013.
Loebenstein, Gad. "Control of Sweet Potato Virus Diseases." *Advances in
    Virus Research* 91 (2015): 33–45. http://doi.org/10.1016/bs.aivir.
    2014.10.005.
Lowder, Sarah K., Jacok Raney, and Terri Raney. "The Number, Size, and
    Distribution of Farms, Smallholder Farms, and Family Farms Worldwide."

World Development 87 (2016): 16–29. http://doi.org/10.1016/j.
worlddev.2015.10.041.

Lumpkin, Thomas, and Janice Armstrong. "Staple Crops, Smallholder
Farmers and Multinationals: The 'Water-Efficient Maize for Africa'
Project." In *World Food Security: Can Private Sector R&D Feed the
Poor?* Crawford Fund for International Agricultural Research, 27–28
October 2009, Parliament House, Canberra, Australia.

Lunduka, Rodney, Monica Fisher, Sieglinde Snapp. "Could Farmer Interest
in a Diversity of Seed Attributes Explain Adoption Plateaus for Modern
Maize Varieties in Malawi?" *Food Policy* 37, no. 5 (2012): 504–10.
http://doi.org/10.1016/j.foodpol.2012.05.001.

Lutaaya, Henry. "Experts Speak Out on Pesticide-Resistant Armyworms."
The Sunrise, 17 March 2017.

Lyatuu, Justus. "Uganda: Pressure to Pass GMO Bill Gains Momentum."
The Observer, 29 March 2017.

Lynas, Mark. "What the Green Movement Got Wrong: A Turncoat
Explains." *The Telegraph*, 4 November 2010.

– "To Abolish Starvation Africa Needs GM Crops." *The Times*, 5 July
2011. https://www.thetimes.co.uk/article/to-abolish-starvation-
africa-needs-gm-crops-3wn6gf3mjdd.

– "Lecture to Oxford Farming Conference, 3 January 2013," 2013. http://
www.marklynas.org/2013/01/lecture-to-oxford-farming-conference-
3-january-2013.

– "Time to Call Out the Anti-GMO Conspiracy Theory." 50th anniversary
celebration of the College of Agriculture and Life Sciences, Cornell
University, 2013. http://www.marklynas.org/2013/04/
time-to-call-out-the-anti-gmo-conspiracy-theory.

– "Tanzania is Burning GM Corn while People Go Hungry." *Little Atoms*,
19 April 2017. http://littleatoms.com/science-world/tanzania-burning-
gm-corn-while-people-go-hungry.

– *Seeds of Science: Why We Got It So Wrong on GMOs.* London:
Bloomsbury, 2018.

– "Seeds of Science: Mark Lynas's U-Turn on Genetically Modified Food."
The Australian, 30 June 2018.

Mabeya, Justin, and Obidimma C. Ezezika. "Unfulfilled Farmer
Expectations: The Case of the Insect Resistant Maize for Africa (IRMA)
Project in Kenya." *Agriculture and Food Security* 1, no. 1 (2012): S6.
http://doi.org/10.1186/2048-7010-1-S1-S6.

Madkour, Magdy A. "Egypt: Biotechnology from Laboratory to the
Marketplace – Challenges and Opportunities." In *Agricultural
Biotechnology and the Poor*, edited by Gabrielle J. Persley, and Manuel

Montecer Lantin, 97–99. Washington, DC: Consultative Group on
International Agricultural Research, 2000.

Madkour, Magdy A., Amin S. El Nawawy, and Patricia L..Traynor.
*Analysis of a National Biosafety System: Regulatory Policies and
Procedures in Egypt.* The Hague: International Service for National
Agricultural Research, 2000.

Martin, Thibaud, Fabrice Chandre, Ochou, Ochou, Maurice Vaissayre,
and Didier Fournier. "Pyrethroid Resistance Mechanisms in the Cotton
Bollworm Helicoverpa Armigera (Lepidoptera: Noctuidae) from West
Africa." *Pesticide Biochemistry and Physiology* 74, no. 1 (2002): 17–26.
http://doi.org/10.1016/S0048-3575(02)00117-7.

Mason, Simon, J. "El Niño, Climate Change, and the Southern African
Climate." *Environmetrics* 12 (2001): 327–45. http://doi.org/10.1002/
env.476.

Matunhu, Jephias. "A Critique of Modernization and Dependency Theories
in Africa: Critical Assessment." *African Journal of History and Culture*
3, no. 5 (2011): 65–72. ISSN 2141-6672.

Mbabazi, Ruth. "Molecular Characterisation and Carotenoid Quantification of Pro-Vitamin A Biofortified Genetically Modified Bananas in
Uganda." PhD dissertation, Queensland University of Technology, 2015.

McCann, James C. *Maize and Grace: Africa's Encounter with a New World
Crop, 1500–2000.* Cambridge, MA: Harvard University Press, 2005.

– "The Political Ecology of Cereal Seed Development in Africa: A History
of Selection." *IDS Bulletin* 42, no. 4 (2011): 24–35. http://doi.
org/10.1111/j.1759-5436.2011.00233.x.

McIntyre, Beverly D., Hans R. Herren, Judi Wakhungu, Robert T. Watson.
Agriculture at a Crossroads: Global Report. Washington, DC: IAASTD,
2009.

McKeon, Nora. "The New Alliance for Food Security and Nutrition: A
Coup for Corporate Capital?" Policy paper, Transnational Institute,
Amsterdam, 2014. https://www.tni.org/files/download/the_new_alliance.
pdf.

McKnight Foundation Collaborative Crop Research Program. "Farmer
Research Networks: Approach Under Development to Build the Evidence
Base for Agroecological Intensification of Smallholder Farming Systems."
N.d. http://www.ccrp.org/sites/default/files/frns-_the_approach.pdf.

McMichael, Philip. "The Land Grab and Corporate Food Regime
Restructuring." *Journal of Peasant Studies* 39, no. 3–4 (2012): 681–701.
http://doi.org/0.1080/03066150.2012.661369.

Meenakshi, Jonnalagadda V., Nancy L. Johnson, Victor M. Manyong,
Hugo DeGroote, Josyline Javelosa, David R. Yanggen, Firdousi Naher,

Carolina Gonzalez, James Garcia, and Erika Meng. "How Cost-Effective is Biofortification in Combating Micronutrient Malnutrition? An Ex Ante Assessment." *World Development* 38, no. 1 (2010): 64–75. http://doi.org/10.1016/j.worlddev.2009.03.014.

Mignouna, Hodeba D., Mathew M. Abang, Gospel Omanya, Francis Nang'Ayo, Mpoko Bokanga, Richard Boadi, Nancy Muchiri, and Eugene Terry. "Delivery of Agricultural Technology to Resource-Poor Farmers in Africa." *Annals of the New York Academy of Sciences* 1,136 (2008): 369–76. http://doi.org/10.1196/annals.1425.010.

Mihale, Matobola Joel, A.L. Deng, H.O. Selemani, M. Mugisha-Kamatenesi, A.W. Kidukuli, and J.O. Ogendo. "Use of Indigenous Knowledge in the Management of Field and Storage Pests Around Lake Victoria Basin in Tanzania." *African Journal of Environmental Science and Technology* 3, no. 9 (2015): 251–9.

Mirondo, Rosemary. "Shock as Government Bans GMO Trials." *The Citizen*, 23 November 2018. https://www.thecitizen.co.tz/News/-Shock-as-government-bans-GMO-trials/1840340-4865040-jp3ji8z/index.html.

Mitchell, Timothy. *Rule of Experts Egypt, Techno-Politics, Modernity.* Berkeley: University of California, 2002.

Monsanto. "Water Efficient Maize for Africa (WEMA)." N.d. https://monsanto.com/company/outreach/water-efficient-maize-africa.

– "User Guide for the Production of YieldGard®, Roundup Ready® Corn 2 and YieldGard® with Roundup Ready® Corn 2." Bryanston, South Africa: Monsanto, 2012. http://www.pannar.com/assets/documents/A09439_Monsanto_User_Guide_Yieldgaurd_II.pdf.

Morris, Jane. "Biosafety Regulatory Systems in Africa," In *Biosafety in Africa: Experiences and Best Practices*, edited by David P. Keetch, Diran Makinde, Cholani K. Weebadde, and Karim M. Maredia. East Lansing, MI: Michigan State University, 2014. http://nepad-abne.net/wp-content/uploads/2015/11/Biosafety-Africa-PartIV2.pdf.

Morse, Stephen, and Robert Bennett. "Impact of Bt Cotton on Farmer Livelihoods in South Africa." *International Journal of Biotechnology* 10, no. 2–3 (2008): 224–39. http://doi.org/10.1504/IJBT.2008.018355.

Morse, Stephen, and A.M. Mannion. "Can Genetically Modified Cotton Contribute to Sustainable Development in Africa?" *Progress in Development Studies* 9, no. 3 (2009): 225–47. http://doi.org/10.1177/146499340800900304.

Moseti, Brian, and Dennis Odunga. "Food Disaster Imminent as Experts Hint at Armyworm Invasion." *Daily Nation*, 13 March 2017. https://www.nation.co.ke/news/Food-disaster-imminent-as-experts-hint-at-army-worm/1056-3847228-ieoooez/index.html.

Mosse, David. *Cultivating Development: An Ethnography of Aid Policy and Practice*. Ann Arbor, MI: Pluto Press, 2005.

Moyin-Jesu, Emmanuel Ibukunoluwa. "Comparative Evaluation of Modified Neem Leaf, Wood Ash and Neem Leaf Extracts for Seed Treatment and Pest Control in Maize (Zea mays L.)." *Emirates Journal of Food and Agriculture* 22, no. 1 (2010): 37–45.

Moyo, Sam. *Land Ownership Patterns and Income Inequality in Southern Africa*. World Economic and Social Survey (WESS), 2015. http://www.un.org/en/development/desa/policy/wess/wess_bg_papers/bp_wess2014_moyo.pdf.

Moyo, Sam, Paris Yeros, and Praveen Jha. "Imperialism and Primitive Accumulation: Notes on the New Scramble for Africa." *Agrarian South: Journal of Political Economy* 1, no. 2 (2013): 181–203. http://doi.org/10.1177/227797601200100203.

Mpinga, James. "State to 'Soften' Bio-Tech Rules." *Daily News*, 27 March 2011. http://gmwatch.eu/latest-listing/1-news-items/13063-farmers-warn-african-governments-over-gm-giants.

Mucioki, Megan, Gordon M. Hickey, Lutta Muhammad, and Timothy Johns. "Supporting Farmer Participation in Formal Seed Systems: Lessons from Tharaka, Kenya." *Development in Practice* 26, no. 2 (2016): 137–48. http://doi.org/10.1080/09614524.2016.1131812.

Mudukuti, Nyasha. "We May Starve, but at Least We'll Be GMO-Free." *Wall Street Journal*, 10 March 2016.

Mudukuti, Nyasha. "Zimbabwe's Rejection of GMO Food Aid Is a Humanitarian Outrage." Global Farmer Network, 14 April 2016. http://globalfarmernetwork.org/2016/04/zimbabwes-rejection-of-gmo-food-aid-is-a-humanitarian-outrage.

Mugisha, Ivan R., and Jean-Pierre Afadhali. "Rwanda Wary of GMOs Use in East Africa." *The East African*, 18 June 2016. http://www.theeastafrican.co.ke/scienceandhealth/Rwanda-wary-of-GMOs-use-in-East-Africa/3073694-3255442-ibf861z/index.html.

Mugo, Stephen, Hugo De Groote, David Bergvinson, Margaret Mulaa, Josephine Songa, and Simon Gichuki. "Developing Bt Maize for Resource-Poor Farmers: Recent Advances in the IRMA Project." *African Journal of Biotechnology* 4, no. 13 (2005): 1490–504. ISSN: 1684-5315.

Mugwagwa, Julius, and Betty Kiplagat. "Legislative and Policy Issues Associated with GE Crops in Africa." In *Biosafety in Africa: Experiences and Best Practice*, edited by David P. Keetch, Diran Makinde, Cholani K. Weebadde, and Karim M. Maredia, 48–53. East Lansing, MI: Michigan State University, 2014.

Muhunyu, Josephat G. "Structural Analysis of Small-Scale Maize Production in the Nakuru District: Challenges Faced in Achieving Stable and High Maize Productivity in Kenya." *Journal of Developments in Sustainable Agriculture* 3 (2008): 74–91.

Mulaa, Margaret, David Bergvinson, Stephen Mugo, Japhether Wanyama, Regina Tende, Hugo Groote, and Tadele Tefera. "Evaluation of Stem Borer Resistance Management Strategies for Bt Maize in Kenya Based on Alternative Host Refugia." *African Journal of Biotechnology* 10, no. 23 (2011): 4732-40. ISSN 1684-5315.

Mulwa, Richard, David Wafula, Margaret Karembu, and Michael Waithaka. "Estimating the Potential Economic Benefits of Adopting Bt Cotton in Selected COMESA Countries," *AgBioForum* 16, no. 1 (2013): 14–26.

Murphy Alexzandra F., Matthew D. Ginzel, and Christian H. Krupke. "Evaluating Western Corn Rootworm (Coleoptera: Chrysomelidae) Emergence and Root Damage in a Seed Mix Refuge." *Journal of Economic Entomology* 103, no. 1 (2010): 147–57. http://doi.org/10.1603/EC09156.

Mutua, Carol. "Watch Out for Banana Disease Giving Farmers Sleepless Nights." *Daily Nation*, 27 May 2016. http://www.nation.co.ke/business/seedsofgold/Malicious-banana-disease-worrying-western-Kenya-banana-farmers-/2301238-3221040-sc1axe/index.html.

Muzaale, Fred. "Small Market for Hybrid Bananas Hurts Farmers." *Daily Monitor*, 5 June 2013. http://www.monitor.co.ug/Magazines/Farming/Small-market-for-hybrid-bananas-hurts-farmers/-/689860/1872932/-/3093n4/-/index.html.

Mwanga, Robert O.M., Charles Niringiye, Agnes Alajo, Benjamin Kigozi, Joweria Namukula, Isaac Mpembe, Silver Tumwegamire, Richard W. Gibson, and Craig G. Yencho. "'NASPOT 11', a Sweetpotato Cultivar Bred by a Participatory Plant-Breeding Approach in Uganda." *HortScience* 46 (2011): 317-21.

Mwanga, Robert O. M., Benson Odongo, Charles Niringiye, Agnes Alajo, Benjamin Kigozi, Rose Makumbi, Esther Lugwana, Joweria Namukula, Isaac Mpembe, Regina Kapinga, Berga Lemaga, James Nsumba, Silver Tumwegamire, and Craig G. Yencho. "'NASPOT 7', 'NASPOT 8', 'NASPOT 9 O', 'NASPOT 10 O', and 'Dimbuka-Bukulula' Sweetpotato." *HortScence* 44 (2009): 828–32. ISSN: 0018-5345.

Myhr, Anne I., and David Quist. *Impact Assessment of Maize MON 87460 from Monsanto*. Norway: GenØk – Centre for Biosafety, 2010.

Nagarajan, Latha, Anwar Naseem, and Carl Pray. "The Political Economy of Genetically Modified Maize in Kenya." *AgBioForum* 19, no. 2 (2016): 198–214.

Nair, Madhavan K., Little Flower Augustine, and Archana Konapur. "Food-Based Interventions to Modify Diet Quality and Diversity to Address Multiple Micronutrient Deficiency." *Frontiers in Public Health* 3 (2015): 277. http://doi.org/10.3389/fpubh.2015.00277.

Namanda, Sam, Rosemary Gatimu, Sam Agili, Stephen Khisa, I. Ndyetabula, and C. Bagambisa. *Micropropagation and Hardening Sweetpotato Tissue Culture Plantlets: A Manual Developed from the SASHA Project's Experience in Tanzania*. Lima, Peru: International Potato Center, 2015.

Nang'Ayo, Francis, Stella Simiyu-Wafukho, and Sylvester O Oikeh. "Regulatory Challenges for GM Crops in Developing Economies: The African Experience." *Transgenic Research* 23, no. 6 (2014): 1,049–55. http://doi.org/10.1007/s11248-014-9805-0.

NARO (National Agricultural Research Organisation). "Economic Impact Brief of Different Crops/Crop Diseases That Are Focus of NARO Biotechnology Research Efforts." N.d.

Nasser, Nagib M.A., and Rodomiro Ortiz. "Cassava Improvement: Challenges and Impacts." *Journal of Agricultural Science* 145, no. 2 (2007): 163–71. http://doi.org/10.1017/S0021859606006575.

National Academies of Sciences, Engineering, and Medicine. *Genetically Engineered Crops: Experiences and Prospects*. Washington, DC: National Academies Press, 2016.

Naylor, Rosamond L., Walter P. Falcon, Robert M. Goodman, Molly M. Jahn, Theresa Sengooba, Hailu Tefera, and Rebecca J. Nelson. "Biotechnology in the Developing World: A Case for Increased Investments in Orphan Crops." *Food Policy* 29, no. 1 (2004): 15–44. http://doi.org/10.1016/j.foodpol.2004.01.002.

Ndungo, Vigheri, S. Eden-Green, G. Blomme, J. Crozier, and J.J. Smith. "Presence of Banana Xanthomonas Wilt (*Xanthomonas campestris* pv. *musacearum*) in the Democratic Republic of Congo (DRC)." *Plant Pathology* 55, no. 2 (2006): 294. http://doi.org/10.1111/j.1365-3059. 2005.01258.x.

Nelson, Godfrey, R. "Case Studies of African Agricultural Biotechnology Regulation: Precautionary and Harmonized Policy-Making in the Wake of the Cartagena Protocol and the AU Model Law." *Loyola of Los Angeles International and Comparative Law Review* 35, no. 3 (2013): 409–32. ISSN: 1533-5860.

Nelson, Rebecca, Richard Coe, and Bettina Haussmann. "Farmer Research Networks as a Strategy for Matching Diverse Options and Contexts in Smallholder Agriculture." *Experimental Agriculture* 1 (2016): 1–20 http://doi.org/10.1017/S0014479716000454.

NEPAD (New Partnership for Africa's Development). "Comprehensive Africa Agriculture Development Programme (CAADP)." N.d. http://www.nepad.org/programme/comprehensive-africa-agriculture-development-programme-caadp.

Netting, Robert M. *Smallholders, Householders: Farm Families and the Ecology of Intensive, Sustainable Agriculture.* Stanford, CA: Stanford University Press, 1993.

New Alliance (New Alliance for Food Security and Nutrition). "About." N.d. https://new-alliance.org/about.

New Scientist. "Monsanto Failure." 181, no. 2,433 (2004): 7. https://www.newscientist.com/article/mg18124330-700-monsanto-failure.

– "Zimbabwe: Govt Says No to GMO Imports Despite 3 Million in Need of Food Assistance." 10 February 2016. http://allafrica.com/stories/201602110188.html.

Newell, Peter. "Bio-Hegemony: The Political Economy of Agricultural Biotechnology in Argentina." *Journal of Latin American Studies* 41, no. 1 (2009): 27–57. http://doi.org/10.1017/S0022216X08005105.

News Day. "Zim Gets $8,1m Food Aid Emergency Fund." 20 November 2015. https://www.newsday.co.zw/2015/11/20/zim-gets-81m-food-aid-emergency-fund.

– "Africa Takes Fresh Look at GMO Crops as Drought Blights Continent." 8 January 2016. https://www.newsday.co.zw/2016/01/08/africa-takes-fresh-look-at-gmo-crops-as-drought-blights-continent.

Ngobese, Thandeka. "Cotton Project Set to Transform Rural Community." *Vuk'uzenzele*, October 2015. http://www.vukuzenzele.gov.za/cotton-project-set-transform-rural-community.

Njuguna, M.M., F.M. Wambugu, S.S. Acharya, and M.A. Mackey. "Socio-Economic Impact of Tissue Culture Banana (*Musa* spp.) in Kenya through the Whole Value Chain Approach." *Acta Horticulturae* 8,791 (2010): 77–86. http://doi.org/10.17660/ActaHortic.2010.879.5.

Nowakunda, Kephas, Patrick R. Rubaihayo, Michael A. Ameny, and Wilberforce Tushemereirwe. "Consumer Acceptability of Introduced Bananas in Uganda." *Infomusa* 9, no. 2 (2000): 22–5.

Ntuli, Nokuthula. "Makhathini Cotton Comeback." *The Mercury*, 18 August 2015. http://www.pressreader.com/south-africa/the-mercury/20150818/281642483909699/TextView.

Nyantakyi-Frimpong, Hanson, and Rachel Bezner Kerr. "A Political Ecology of High-Input Agriculture in Northern Ghana." *African Geographical Review* 34, no. 1 (2015): 13–35. http://doi.org/10.1080/19376812.2014.929971.

Oakland Institute. *Farmer Participatory Research (FPR) in Uganda.*
Agroecology Case Studies. Oakland, CA: Oakland Institute, 2015.
https://www.oaklandinstitute.org/sites/oaklandinstitute.org/files/
Participatory_Research_Uganda.pdf.

Oben, Esther Obi, Nelson Neba Ntonifor, Sevilor Kekeunou, and Martin
Nkwa Abbeytakor. "Farmers' Knowledge and Perception on Maize
Stem Borers and their Indigenous Control Methods in South Western
Region of Cameroon." *Journal of Ethnobiology and Ethnomedicine* 11,
no. 77 (2015). http://doi.org/10.1186/s13002-015-0061-z.

Observer. "Parliament Passes GMO Bill." 29 November 2018. https://www.
observer.ug/businessnews/59356-parliament-passes-gmo-bill.

Ochieng, Justus, and Faith Matete. "Kenya: Governors Urge State to Lift
Ban on GMO Foods." *AllAfrica*, 17 July 2014. http://allafrica.com/
stories/201407171627.html.

Odame, Hannington, Patricia Kameri-Mbote, and David Wafula. "Innovation and Policy Process: Case of Transgenic Sweet Potato in Kenya."
Economic and Political Weekly 37, no. 27 (2002): 2770-7. ISSN:
00129976.

– *Governing Modern Agricultural Biotechnology in Kenya: Implications
for Food Security.* Brighten, UK: Institute of Development Studies, 2003.

Odhiambo, Judith A., Urszula Norton, Dennis Ashilenje, Emmanuel Omondi,
and Jay Norton. "Weed Dynamics during Transition to Conservation
Agriculture in Western Kenya Maize Production." *PloS One* 10, no. 8
(2015): E0133976. http://doi.org/10.1371/journal.pone.0133976.

Odhiambo, Moses. "Governors: Lift Ban on Genetic Foods." *Daily
Nation*, 15 July 2014. http://www.nation.co.ke/counties/kisumu/
Governors-Genetically-Modified-Crops-Food-Shortages-GMOs/-
/1954182/2385374/-/2x50xu/-/index.html.

OFAB (Open Forum on Agricultural Biotechnology) Tanzania. "Tanzania:
GMOs Good for Agriculture Development, Says Don." OFAB, 19 July
2013. http://ofabtanzania.blogspot.ca/2013/07/tanzania-gmos-good-for-
agriculture.html.

Oikeh, Sylvester, Dianah Ngonyamo-Majee, Stephen I.N. Mugo,
Kingstone Mashingaidze, Vanessa Cook, and Michael Stephens. "The
Water Efficient Maize for Africa Project as an Example of a Public-
Private Partnership" In *Convergence of Food Security, Energy Security
and Sustainable Agriculture*, edited by David D. Songstad, Jerry L.
Hatfield, and D.T. Tomes, 317-29. Heidelberg: Springer, 2014.

Ojanji, Wandera. "Partners Recognize Achievements in Insect-Resistant
Maize Delivery." CIMMYT, 21 April 2014. http://www.cimmyt.org/
partners-recognize-achievements-in-insect-resistant-maize-delivery.

Okine, Charles Benoni. "Trial of New Cotton Variety Positive in 3 Northern Regions; But No Commercialisation Yet." *Graphic Online*, 29 June 2015. http://www.graphic.com.gh/business/business-news/trial-of-new-cotton-variety-positive-in-3-northern-regions-but-no-commercialisation-yet.html.

Okogbenin, Emmanuel, Chiedozie Egesi, and Martin Fregene. "Molecular Markers and Tissue Culture: Technologies Transcending Continental Barriers to Add Value and Improve Productivity of Cassava in Africa." In *Biotechnologies at Work for Smallholders: Case Studies from Developing Countries in Crops, Livestock and Fish*, edited by John Ruane, James D. Dargie, C. Mba, Paul John Boettcher, Harinder Makkar, Devin Barthey, and Andrea Sonnino, 37–46. Rome: Food and Agriculture Organization of the United Nations (FAO), 2013.

Olsen, Kelby. "Monsanto in Egypt: Activists Refuse to be Human Experiments." *Muftah*, 30 May 2016. http://muftah.org/monsanto-in-egypt-activists-refuse-to-be-human-experiments/#.VUfHI2byo2f.

Onduru, Davies D., P. Snijders, F.N. Muchena, B. Wouters, A. De Jager, L. Gachimbi, and G.N. Gachini. "Manure and Soil Fertility Management in Sub-Humid and Semi-Arid Farming Systems of Sub-Saharan Africa: Experiences from Kenya." *International Journal of Agricultural Research* 3 (2008): 166–87. http://doi.org/10.3923/ijar.2008.166.187.

Ongu, Isaac. "Does Uganda Need GMOs? Scientists Look to Gene Editing to Spur Innovation." *Genetic Literacy Project*, 5 September 2017. https://geneticliteracyproject.org/2017/09/05/uganda-need-gmos-scientists-look-gene-editing-spur-innovation.

Onstad, David W., Paul D. Mitchell, Terrance M. Hurley, Jonathan G. Lundgren, R. Patrick Porter, Christian H. Krupke, Joseph L. Spencer, Christine D. Difonzo, Tracey S. Baute, Richard L. Hellmich, Lawrent L. Buschman, William D. Hutchison, and John F. Tooker. "Seeds of Change: Corn Seed Mixtures for Resistance Management and Integrated Pest Management." *Journal of Economic Entomology* 104, no. 2 (2011): 343–52. ISSN: 0022-0493.

Otieno, Rawlings. "Lift Genetically Modified Organisms Ban for Food Security, State Told." *Standard Digital*, 30 May 2013. http://www.standardmedia.co.ke/article/2000122945/lift-genetically-modified-organisms-ban-for-food-security-state-told.

Oxfam. "Moral Hazard? 'Mega' Public-Private Partnerships in African Agriculture." Oxfam briefing paper no. 188. Oxfam: Oxford, UK, 2014. https://www.oxfam.org/sites/www.oxfam.org/files/file_attachments/oxfam_moral_hazard_ppp-agriculture-africa-010914-embargo-en.pdf.

Paarlberg, Robert. "Toward a Regional Policy on GMO Crops among COMES/ASARECA Countries." RABESA regional workshop policy options paper, Nairobi: African Center for Technology Studies, 2006.

– "A Dubious Success: The NGO Campaign against GMOs." *GM Crops and Food* 5, no. 3 (2014): 223–28. http://doi.org/10.4161/21645698.20 14.952204.

Patel, Raj. "The Long Green Revolution." *Journal of Peasant Studies* 40, no. 1 (2013): 1–63. http://doi.org/10.1080/03066150.2012.719224.

PBS (Program for Biosafety Systems). "About Us." N.d. http://pbs.ifpri. info/about.

Pehu, Eija, and Catherine Ragasa. *Agricultural Biotechnology: Transgenics in Agriculture and Their Implications for Developing Countries.* Washington, DC: World Bank, 2008.

Perreault, Tom, Gavin Bridge, and James McCarthy. *Routledge Handbook of Political Ecology.* New York: Routledge, 2015.

Pesa Times. 2013. "GMOs in Tanzania, Scientist Wants Review." 6 March. http://pesatimes.co.tz/news/agriculture/gmos-in-tanzania-scientist-wants-review.

Peterman, Amber, Julia Behrman, and Agnes Quisumbing. "A Review of Empirical Evidence on Gender Differences in Nonland Agricultural Inputs, Technology, and Services." IFPRI discussion paper 00975. International Food Policy Research Institute, Washington, DC, 2010.

Petit, Michel. *The Benefits of Modern Agriculture: A Reassessment Following Recent Controversies.* Pre-publication draft. Washington, DC: Global Harvest Initiative, 2010.

Pingali, Prabhu. "Green Revolution: Impacts, Limits, and the Path Ahead." *Proceedings of the National Academy of Sciences of the United States of America* 109, no. 31 (2012): 12302–28. http://doi.org/10.1073/pnas. 0912953109.

Poulton, Colin. "The Cotton Sector in Northern Ghana: Liberalisation, Seasonal Input Use and Impact of Structural Adjustment." In *Natural Resource Management in Ghana and its Socio-Economic Context*, edited by Roger Blench, 70–84. London: Overseas Development Institute, 1999.

Poulton, Colin, Peter Gibbon, Benjamine Hanyani-Mlambo, Jonathan Kydd, Wilbald Maro, Marianne Nylandsted Larsen, Afonso Osorio, David Tschirley, and Ballard Zulu. "Competition and Coordination in Liberalized African Cotton Market Systems." *World Development* 32, no. 3 (2004): 519–36. http://doi.org/10.1016/j.worlddev. 2003.10.003.

Preston-Whyte, Rob A., and Peter Daughtrey Tyson. *The Atmosphere and Weather of Southern Africa*. Cape Town, South Africa: Oxford University Press, 1988.

Pretty, Jules, Camilla Toulmin, and Stella Williams. "Sustainable Intensification in African Agriculture." *International Journal of Agricultural Sustainability* 9, no. 1 (2011): 5–24. http://doi.org/10.3763/ijas.2010.0583.

Qaim, Matin. *The Economic Effects of Genetically Modified Orphan Commodities: Projections for Sweet Potato in Kenya*. Ithaca, NY: ISAAA, 1999.

– "Benefits of Genetically Modified Crops for the Poor: Household Income, Nutrition, and Health." *New Biotechnology* 27, no. 5 (2010): 552–7. http://doi.org/10.1016/j.nbt.2010.07.009.

– *Genetically Modified Crops and Agricultural Development*. New York: Palgrave Macmillan, 2016.

Ranneberger, Michael. "Cautious Kenya Finally Enacts Long-Awaited Biosafety Act of 2009." Wikileaks Cables: 09NAIROBI496. Dated 11 March 2009.

Reeder, R.H., John Baptist Muhinyuza, O. Opolot, V. Aritua, Jayne Crozier, and Julian Smith. "Presence of Banana Bacterial Wilt (*Xanthomonas campestris* pv. *musacearum*) in Rwanda." *Plant Pathology* 56, no. 6 (2007): 1,038. http://doi.org/10.1111/j.1365-3059.2007.01640.x.

Regier, Gregory K., Timothy J. Dalton, and Jeffery R. Williams. "Impact of Genetically Modified Maize on Smallholder Risk in South Africa." *AgBioForum* 15, no. 3 (2012): 328–36.

Resnik, David. "Retracting Inconclusive Research: Lessons from the Séralini GM Maize Feeding Study." *Journal of Agricultural and Environmental Ethics* 28, no. 4 (2015): 621–33. http://doi.org/10.1007/s10806-015-9546-y.

Rietveld, Anne, Susan Ajambo, and Enoch Kikulwe. "Economic Gain and Other Losses? Gender Relations and Matooke Production in Western Uganda." Conference on International Research on Food Security, Natural Resource Management and Rural Development, Vienna, Austria, 18–21 September 2016.

Rietveld, Anne, Wellington Jogo, Samuel Mpiira, and Charles Staver. "The Effect of Banana Xanthomonas Wilt on Beer-banana Value Chains in Central Uganda: An Exploratory Study." *Journal of Agribusiness in Developing and Emerging Economies* 4, no. 2 (2014): 172–84. http://doi.org/10.1108/JADEE-08-2012-0021.

Rijssen, F.W. Jansen van, Elizabeth J. Morris, Jacobus N. Eloff. "A Critical Scientific Review on South African Governance of Genetically Modified Organisms." *African Journal of Biotechnology* 12, no. 32 (2013): 5010–21. http://doi.org/10.5897/AJB12.2158.

Rivera, Ana Leonor, Miguel Gómez-Lim, Francisco Fernández, and Achim M. Loske. "Physical Methods for Genetic Plant Transformation." *Physics of Life Reviews* 9, no. 3 (2012): 308–45. http://doi.org/10.1016/j.plrev.2012.06.002.

Robbins, Paul. *Political Ecology: A Critical Introduction – Critical Introductions to Geography*. Malden, MA: Blackwell, 2004.

Roberts, Richard. *Two Worlds of Cotton: Colonialism and the Regional Economy in the French Soudan 1800–1946*. Stanford, CA: Stanford University Press, 1996.

Ronald, Pamela C. "Lab to Farm: Applying Research on Plant Genetics and Genomics to Crop Improvement." *PLoS Biology* 12, no. 6 (2014): e1001878. http://doi.org/10.1371/journal.pbio.1001878.

Rooney, William L. "Sorghum Improvement: Integrating Traditional and New Technology to Produce Improved Genotypes." *Advances in Agronomy* 83 (2004): 37–109. http://doi.org/10.1016/S0065-2113(04)83002-5.

Ruane, John, James D. Dargie, Chikelu Mba, Paul Boettcher, Harinder Makkar, Devin M. Bartley, and Andrea Sonnino. *Biotechnologies at Work for Smallholders: Case Studies from Developing Countries in Crops, Livestock and Fish*. Rome: Food and Agriculture Organization of the United Nations (FAO), 2013.

Rudi, Nderim, George W. Norton, Jeffrey Roger Alwang, and Godwin N. Asumugha. "Economic Impact Analysis of Marker-Assisted Breeding for Resistance to Pests and Post Harvest Deterioration in Cassava." *African Journal of Agricultural Resource Economics* 4, no. 2 (2010): 110–22. http://doi.org/10.2307/40588528.

Rwegasira, Gration M. "Prevalence of Sweetpotato Virus Vectors in Lake Victoria Basin of Tanzania: A Challenge to Disease Management Techniques." *Journal of Agricultural Science and Technology* 1, no. 11 (2011): 979–86. http://doi.org/10.17265/2161-6256/2011.11A.005.

Sarant, Louise. "Egypt's Legal Battle to Regulate Monsanto's GMOs." *Egypt Independent*, 21 July 2012. http://www.egyptindependent.com/news/egypt-s-legal-battle-regulate-monsanto-s-gmos-o.

Schimmelpfennig, David, Stacey Rosen, and Carl Pray. *Genetically Engineered Corn in South Africa: Implications for Food Security in the*

Region. Economic Research Service, United States Department of Agriculture, Washington, DC, 2013.

Schmickle, Sharon. "Tanzania Becomes a Battleground in Fight Over Genetically Modified Crops." *Washington Post*, 7 October 2013.

Schnurr, Matthew A. "Lowveld Cotton: A Political Ecology of Agricultural Failure in Natal and Zululand, 1844–1948." PhD dissertation, University of British Columbia, 2008.

– "Breeding for Insect-Resistant Cotton across Imperial Networks, 1924–1950." *Journal of Historical Geography* 37, no. 2 (2011): 223–31. http://doi.org/10.1016/j.jhg.2010.09.002.

– "Inventing Makhathini: Creating a Prototype for the Dissemination of Genetically Modified Crops into Africa." *Geoforum* 43, no. 4 (2012): 784–92. http://doi.org/10.1016/j.geoforum.2012.01.005.

– "Biotechnology and Bio-Hegemony in Uganda: Unraveling the Social Relations Underpinning the Promotion of Genetically Modified Crops into New African Markets." *Journal of Peasant Studies* 40, no. 4 (2013): 639–58.

– "GMO 2.0: Genetically Modified Crops and the Push for Africa's Green Revolution." *Canadian Food Studies* 2, no. 2 (2015): 201–8. http://doi.org/10.15353/cfs-rcea.v2i2.97.

Schnurr, Matthew A., and Lincoln Addison. "Which Variables Influence Farmer Adoption of Genetically Modified Orphan Crops? Measuring Attitudes and Intentions to Adopt GM Matooke Banana in Uganda." *AgBioForm* 20, no. 2 (2017): 133–47.

Schnurr, Matthew A., and Christopher Gore. "Getting to 'Yes': Governing Genetically Modified Crops in Uganda." *Journal of International Development* 27, no. 1 (2015): 55–72. http://doi.org/10.1002/jid.3027.

Schnurr, Matthew A., Lincoln Addison, and Sarah Mujabi-Mujuzi. "Limits to Biofortification: Farmer Perspectives on a Vitamin A Enriched Banana in Uganda." *Journal of Peasant Studies*, 2018. http://doi.org/10.1080/03066150.2018.1534834.

Schroeder, Richard A. "Shady Practice: Gender and the Political Ecology of Resource Stabilization in Gambian Garden/Orchards." *Economic Geography* 69, no. 4 (1993): 349–65. http://doi.org/10.2307/143594.

Schurman, Rachel. "Building an Alliance for Biotechnology in Africa." *Journal of Agrarian Change* 17, no. 3 (2017): 441–58.

Schwartz, Alfred. "L'évolution de l'agriculture en zone cotonnière dans l'Ouest du Burkina Faso." In *Défis Agricoles Africains*, edited by Jean-Claude Devèze, 153–72. Paris: Karthala, 2007.

Scoones, Ian, and John Thompson. *Farmer First Revisited: Innovation for Agricultural Research and Development*. Worwideshire, UK: Practical Action, 2009.

– "The Politics of Seed in Africa's Green Revolution: Alternative Narratives and Competing Pathways." *IDS Bulletin* 42, no. 4 (2011): 1–23. http://doi.org/10.1111/j.1759-5436.2011.00232.x.

Serraj, Rachid, C. Tom Hash, S. Masood H. Rizvi, Arun Sharma, Rattan S. Yadav, and Fran R. Bidinger. "Recent Advances in Marker-assisted Selection for Drought Tolerance in Pearl Millet." *Plant Production Science* 3 (2005): 334–7. http://doi.org/10.1626/pps.8.334.

Shanghai Daily. "Zambia Allows South African Chain Store to Import GMO Foods." *Shanghai Daily*, 20 April 2016. http://www.shanghaidaily.com/article/article_xinhua.aspx?id=327321.

Shankar, Bhavani, Richard Bennett, and Steve Morse. "Output Risk Aspects of Genetically Modified Crop Technology in South Africa." *Economics of Innovation and New Technology* 16, no. 4 (2007): 277–91. http://doi.org/10.1080/10438590600692926.

Sheridan, Michael J. "An Irrigation Intake Is like a Uterus: Culture and Agriculture in Precolonial North Pare, Tanzania." *American Anthropologist* 104, no. 1 (2002): 79–92. http://doi.org/10.1525/aa.2002.104.1.79.

Shiferaw, Bekele, Boddupalli M. Prasanna, Jonathan Hellin, and Marianna Bänziger. "Crops That Feed the World: 6. Past Successes and Future Challenges to the Role Played by Maize in Global Food Security." *Food Security* 3, no. 3 (2011): 307–27. http://doi.org/10.1007/s12571-011-0140-5.

Shimwela, Mpoki M., Jason K. Blackburn, Jeff B. Jones, J. Nkuba, Hossein A. Narouei-Khandan, Randy C. Ploetz, Fenton Beed, and Ariena van Bruggen. "Local and Regional Spread of Banana Xanthomonas Wilt (BXW) in Space and Time in Kagera, Tanzania." *Plant Pathology* 66, no. 6 (2017): 1003–14. http://doi.org/10.1111/ppa.12637.

Singh, R.B. "Environmental Consequences of Agricultural Development: A Case Study from the Green Revolution State of Haryana, India." *Agriculture, Ecosystems and Environment* 82, no. 1 (2000): 97–103. http://doi.org/10.1016/S0167-8809(00)00219-X.

Smale, Melinda. "GMOs and Poverty: The Relationship Between Improved Seeds and Rural Transformations." *Canadian Journal of Development Studies* 38, no. 1 (2017): 139–48. http://doi.org/10.1080/02255189.2016.1208607.

Smale, Melinda, and Thom Jayne. "Maize in Eastern and Southern Africa: 'Seeds' of Success in Retrospect." EPTD discussion paper no. 97,

International Food Policy Research Institute, Washington, DC, 2003. http://www.fao.org/docs/eims/upload/166420/Smale,Jayne.pdf.

Smale, Melinda, Derek Byerlee, and Thom Jayne. "Maize Revolutions in Sub-Saharan Africa." Policy research working paper 5659, World Bank, Washington, DC, 2011.

Smale, Melinda, Paul W. Heisey, and Howard D. Leathers. "Maize of the Ancestors and Modern Varieties: The Microeconomics of High-Yielding Variety Adoption in Malawi." *Economic Development and Cultural Change* 43, no. 2 (1995): 351–68. http://doi.org/10.1086/452154.

Sobha, I. "Green Revolution: Impact on Gender." *Journal of Human Ecology* 22, no. 2 (2007): 107–13. http://doi.org/10.1080/09709274.20 07.11906008.

Soleri, Daniela, David A. Cleveland, Flavio A. Cuevas, Mario R. Fuentes, Humberto L. Ríos, and Stuart H. Sweeney. "Understanding the Potential Impact of Transgenic Crops in Traditional Agriculture: Maize Farmers' Perspectives in Cuba, Guatemala and Mexico." *Environmental Biosafety Research* 4 (2005): 141–66. http://doi.org/10.1051/ ebr:2005019.

Soleri, Daniela, David A. Cleveland, Garrett Glasgow, Stuart H. Sweeney, Flavio A. Cuevas, Mario R. Fuentes, and Humberto L. Ríos. "Testing Assumptions Underlying Economic Research on Transgenic Food Crops for Third World Farmers: Evidence from Cuba, Guatemala and Mexico." *Ecological Economics* 67, no. 4 (2008): 667–82. http://doi. org/10.1016/j.ecolecon.2008.01.031.

Solomon, Salem. "Farmers Search for Answers as Armyworms Move Across Africa." *VA ONEWS*, 12 May 2017. https://www.voanews.com/a/ farmers-africa-armyworms-destroy-crops/3849723.html.

Southern African Regional Biosafety Network. "Media Has Key Role in African Biosafety," *AgBioView*, 15 November 2001.

Spielman, David J., and Patricia Zambrano. "Policy, Investment, and Part-nerships for Agricultural Biotechnology Research in Africa: Emerging Evi-dence." In *Genetically Modified Crops in Africa: Economic and Policy Lessons from Countries South of the Sahara*, edited by Falck-Zepeda, José Benjamin, Guillaume P. Gruère, and Idah Sithole-Niang, 183–206. Washington, DC: International Food Policy Research Institute, 2013.

Spielman, David J., Frank Hartwich, and Klaus Grebmer. "Public-Private Partnerships and Developing-Country Agriculture: Evidence from the International Agricultural Research System." *Public Administration and Development* 30, no. 4 (2010): 261–76. http://doi.org/10.1002/pad.574.

Ssonko, Sunrise. "Monsanto Chief Points at Roadblocks Which Failed Uganda's BT Cotton Trial." *The Sunrise*, 18 December 2013.

– "ActionAid Admits to Misleading Ugandans Over GMOs." *The Sunrise*, 27 March 2015.

Stads, Gert-Jan, and Nienke Beintema. "Agricultural R&D Expenditure in Africa: An Analysis of Growth and Volatility." *European Journal of Development Research* 27, no. 3 (2015): 391–406. http://doi.org/ 10.1057/ejdr.2015.25.

Stieber, Zachary. "GMOs, A Global Debate: Zambia, Strongest Anti-GMO Stance in Africa." *Epoch Times*, 7 September 2013.

Stone, Glenn Davis. "Agricultural Deskilling and the Spread of Genetically Modified Cotton in Warangal." *Current Anthropology* 48, no. 1 (2007): 67–103. http://doi.org/10.1086/508689.

– "Constructing Facts: Bt Cotton in India." *Economic and Political Weekly* 47, no. 38 (2012): 62–70. ISSN: 00129976.

– "Towards a General Theory of Agricultural Knowledge Production: Environmental, Social, and Didactic Learning." *Culture, Agriculture, Food and Environment* 38, no. 1 (2016): 5–17. http://doi.org/10.1111/ cuag.12061.

Storer, Nicholas P., Mary E. Kubiszak, J. Ed King, Gary D. Thompson, and Antonio Cesar Santos. "Status of Resistance to Bt Maize in Spodoptera Frugiperda: Lessons from Puerto Rico." *Journal of Invertebrate Pathology* 110, no. 3 (2012): 294–300. http://doi.org/10.1016/j. jip.2012.04.007.

Suliman, Elnayer H., Osama M.A. Elhassan, Abdelbagi M. Ali, and Nasrein M. Kamal. "Evaluation of Two Chinese Bt Cotton Genotypes Against the African Bollworm, Helicoverpa Armigera (Hubner) Damage Under Sudan Rainfed Conditions." *Scholars Journal of Agriculture and Veterinary Sciences* 2, no. 2A (2015): 102–7.

Sulle, Emmanuel, and Ruth Hall. "Reframing the New Alliance Agenda: A Critical Assessment Based on Insights from Tanzania." FAC policy brief 56, Brighton, UK: Future Agricultures Consortium, 2013. http://www. plaas.org.za/plaas-publication/fac-pb56.

Sumberg, James. "Could the 3Ds Breathe New Life into Farming Systems Research?" STEPS Centre: Pathways to Sustainability, 30 April 2013. https://steps-centre.org/blog/could-the-3ds-breathe-new-life-into- farming-systems-research.

– *Agronomy for Development: The Politics of Knowledge in Agricultural Research*. New York: Routledge, 2017.

Suri, Shivali, Dinesh Kumar, and Ranjan Das. "Dietary Deficiency of Vitamin A among Rural Children: A Community-Based Survey Using a Food-Frequency Questionnaire." *National Medical Journal of India* 30, no. 2 (2017): 61–4. ISSN: 0970-258X.

Syngenta Foundation. "Insect Resistant Maize for Africa – IRMA." N.d. https://www.syngentafoundation.org/insect-resistant-maize-africa-irma.

Tabashnik, Bruce E., Thierry Brévault, and Yves Carrière. "Insect Resistance to Bt Crops: Lessons from the First Billion Acres." *Nature Biotechnology* 31, no. 6 (2013): 510–21. http://doi.org/10.1038/nbt.2597.

Tabashnik, Bruce E., J. B. J. Van Rensburg, and Yves Carrière. "Field-Evolved Insect Resistance to Bt Crops: Definition, Theory, and Data." *Journal of Economic Entomology* 102 (2009): 2011–25. http://doi.org/10.1603/029.102.0601.

Tabashnik, Bruce E., David Mota-Sanchez, Mark E. Whalon, Robert M. Hollingworth, and Yves Carrière. "Defining Terms for Proactive Management of Resistance to Bt Crops and Pesticides." *Journal of Economic Entomology* 107, no. 2 (2014): 496–507. http://doi.org/10.1603/EC13458.

Tadele, Zerihun, and Kebebew Assefa. "Increasing Food Production in Africa by Boosting the Productivity of Understudied Crops." *Agronomy* 2, no. 4 (2012): 240–83. http://doi.org/10.3390/agronomy2040240.

Tairo, Fred, Settumba B. Mukasa, Roger A.C. Jones, Alois Kullaya, Patrick R. Rubaihayo, and Jari P.T. Valkonen. "Unravelling the Genetic Diversity of the Three Main Viruses Involved in Sweet Potato Virus Disease (SPVD), and Its Practical Implications." *Molecular Plant Pathology* 6, no. 2 (2005): 199–211. http://doi.org/10.1111/j.1364-3703.2005.00267.x.

Talukder, Aminuzzaman, Lynnda Kiess, Nasreen Huq, Saskia de Pee, Ian Darnton-Hill, and Martin W. Bloem. "Increasing the Production and Consumption of Vitamin A-Rich Fruits and Vegetables: Lessons Learned in Taking the Bangladesh Homestead Gardening Programme to a National Scale." *Food and Nutrition Bulletin* 21, no. 2 (2000): 165–72. http://doi.org/10.1177/156482650002100210.

Tanzania Daily News. "Tanzania: Growing of Biotech Crops Increases, Forum Told," 24 July 2015. http://allafrica.com/stories/201507240398.html.

Tefera, Tadele, Stephen Mugo, Murenga Mwimali, Bruce Anani, Regina Tende, Yoseph Beyene, Simon Gichuki, Sylvester O Oikeh, Francis Nang'Ayo, James Okeno, Evans Njeru, Kiru Pillay, Barbara Meisel, and B.M. Prasanna. "Resistance of Bt-Maize (MON810) against the Stem Borers *Busseola fusca* (Fuller) and *Chilo partellus* (Swinhoe) and Its Yield Performance in Kenya." *Crop Protection* 89 (2016): 202–8. http://doi.org/10.1016/j.cropro.2016.07.023.

Tembo, Benedict. "Genetically Modified Bt Maize Alternative for Combating Armyworms." *Zambia Daily Mail Limited*, 20 February

2017. https://www.daily-mail.co.zm/genetically-modified-bt-maize-alternative-for-combating-armyworms.

Thomson, Jennifer. "South Africa: An Early Adopter of GM Crops." In *Insights: Africa's Future ... Can Biosciences Contribute?*, edited by Brian Heap and David Bennett, 64–9. Cambridge, UK: Branson/B4FA, 2013.

Tilley, Helen. *Africa as a Living Laboratory: Empire, Development, and the Problem of Scientific Knowledge, 1870–1950.* Chicago, IL: University of Chicago Press, 2011.

Traoré, Hamidou, S.A.O. Hema, and Karim Traoré. "Bt Cotton in Burkina Faso Demonstrates That Political Will Is Key for Biotechnology to Benefit Commercial Agriculture in Africa." In *Biotechnology in Africa: Emergence, Initiatives and Future*, edited by Florence Wambuga, Daniel Kamanga, and Ismail Serageldin, 15–36. New York: Springer International, 2014.

Traoré, Oula, Denys Sanfo, Karim Traoré, and Bazoumana Koulibaly. *The Effect of Bt Gene on Cotton Productivity, Ginning Rate and Fiber Characteristics under Burkina Faso Cropping Conditions.* Bobo Dialasso, Burkina Faso: Institut de l'Environnement et de Recherches Agricoles (INERA), 2006.

– "Testing the Efficacy and Economic Potential of Bollgard II under Burkina Faso Cropping Conditions." *Journal of Cotton Science* 12, no. 2 (2008): 87–98. ISSN: 1524-3303.

Tripp, Robert Burnet. *Biotechnology and Agricultural Development Transgenic Cotton, Rural Institutions and Resource-Poor Farmers.* New York: Routledge, 2009.

Tumusiime, Emmanuel, Hugo De Groote, Jeffrey Vitale, and Brian Adam. "The Cost of Coexistence between Bt Maize and Open-Pollinated Maize Varieties in Lowland Coastal Kenya." *AgBioForum* 13, no. 3 (2010): 208–21.

Tushemereirwe, Wilberforce, A. Kangire, J. Smith, M. Nakyanzi, D. Kataama, and C. Musiitwa. *An Outbreak of Banana Bacterial Wilt in Mukono District: A New and Devastating Disease.* National Banana Programme (Uganda), Kawanda Agricultural Research Organisation, 2001.

Tushemereirwe, Wilberforce, A. Kangire, F. Ssekiwoko, L.C. Offord, J. Crozier, E. Boa, M. Rutherford, and J.J. Smith. "First Report of *Xanthomonas campestris* pv. *musacearum* on Banana in Uganda." *Plant Pathology* 53, no. 6 (2004): 802. http://doi.org/10.1111/j.1365-3059.2004.01090.x.

Tushemereirwe, Wilberforce, A. Kangire, J. Smith, F. Ssekiwoko, M. Nakyanzi, D. Kataama, C. Musiitwa, and R. Karyaija. "An Outbreak of Bacterial Wilt on Banana in Uganda." *InfoMusa* 12, no. 2 (2003): 6–8.

Tyson, Peter D., and Thomas G.J. Dyer. "Mean Annual Fluctuations of Precipitation in the Summer Rainfall Region of South Africa." *South African Geographical Journal* 57, no. 2 (1975): 105–10. http://doi.org/1 0.1080/03736245.1974.10559553.

Uganda Biosciences Information Center. *Facts and Figures: Major Crops under Biotech Research in Uganda.* Entebbe: National Agricultural Research Organisation, 2015.

Uganda Bureau of Statistics (UBOS) and ICF International. "Uganda: Demographic and Health Survey 2011," 2012. http://dhsprogram.com/ pubs/pdf/FR264/FR264.pdf.

UN DESA (United Nations Department of Economic and Social Affairs) Population Division. *World Population Prospects: The 2015 Revision – Key Findings and Advance Tables.* Washington, DC: United Nations, 2015.

UNHRC (United Nations Human Rights Council). "Report submitted by the Special Rapporteur on the Right to Food, Olivier De Schutter," 20 December 2010. UN General Assembly. A/HRC/16/49. http://www. srfood.org/images/stories/pdf/officialreports/20110308_a-hrc-16-49_ agroecology_en.pdf.

United Republic of Tanzania. Environmental Management Act, Environmental Management (Biosafety) Regulations, 2009. Article 6. http://tz.chm-cbd.net/biosafety/national-implementation/national- biosafety-framework/tanzania-biosafety-regulations-2009.pdf.

USAID (United States Agency for International Development). "More than $3 billion in Private Sector Investment for the New Alliance for Food Security and Nutrition," 2013. https://www.usaid.gov/news-information/ fact-sheets/more-3-billion-private-sector-investment-new-alliance- food-security-and.

– "Agriculture and Food Security," 31 July 2018. https://www.usaid.gov/ what-we-do/agriculture-and-food-security.

USDA (United States Department of Agriculture). *Agricultural Statistics 1960.* Washington, DC: United States Government Printing Office, 1961s.

– "Monsanto Company Petition (07-CR-191U) for Determination of Non-Regulated Status of Event MON 87460: Final Environmental Assessment," 2011.

– *Egypt: Agricultural Biotechnology Annual 2014.* Global Agricultural Information Network, Foreign Agricultural Service, 2014. http://gain.

fas.usda.gov/Recent%20GAIN%20Publications/Agricultural%20
Biotechnology%20Annual_Cairo_Egypt_7-10-2014.pdf.

– "Ghana Cotton Production by Year," 2018. http://www.indexmundi.
com/agriculture/?country=gh&commodity=cotton&graph=production.

USSEC (United States Soybean Export Council). "Kenyan Governors Call
for the Lifting of GMO Ban," 2 June 2014. http://ussec.org/kenyan-
governors-call-lifting-gmo-ban.

Uys, Gerhard. "GM Maize Could Be Less Susceptible to Fall Armyworm."
Farmer's Weekly, 13 February 2017. http://www.farmersweekly.co.za/
agri-news/south-africa/gm-maize-less-susceptible-fall-armyworm.

Van Beusekom, Monica M. *Negotiating Development: African Farmers
and Colonial Experts at the Office du Niger, 1920–1960*. Portsmouth,
NH: Heinemann, 2002.

Van den Berg, Johnnie. "Insect Resistance Management in Bt Maize: Wild
Host Plants of Stem Borers Do Not Serve as Refuges in Africa." *Journal
of Economic Entomology* 110, no. 1 (2017): 221–9. http://doi.org/
10.1093/jee/tow276.

Van Den Berg, Johnniem, Angelika Hilbeck, and Thomas Bøhn. "Pest
Resistance to Cry1Ab Bt Maize: Field Resistance, Contributing Factors
and Lessons from South Africa." *Crop Protection* 54 (2013): 154–60.
http://doi.org/10.1016/j.cropro.2013.08.010.

Van Der Straeten, Dominique, Teresa B. Fitzpatrick, and Hans De Steur.
"Editorial Overview: Biofortification of Crops: Achievements, Future
Challenges, Socio-Economic, Health and Ethical Aspects." *Current
Opinion in Biotechnology* 44 (2017): 7–10. http://doi.org/10.1016/j.
copbio.2017.03.007.

Van Rensburg, J.B.J. "First Report of Field Resistance by the Stem Borer,
Busseola fusca (Fuller) to Bt-Transgenic Maize." *South African Journal
of Plant and Soil* 24 (2007): 147–51. http://doi.org/10.1080/02571862.
2007.10634798.

Van Rensburg, J.B.J., J.H. Giliomee, and M.C. Walters. "The Influence of
Rainfall on the Seasonal Abundance and Flight Activity of the Maize
Stalk Borer, Busseola Fusca in South Africa." *South African Journal of
Plant and Soil* 4 (1987): 183–7. http://doi.org//10.1080/02571862.1987.
10634971.

Vanloqueren, Gaëtan, and Philippe V. Baret. "How Agricultural Research
Systems Shape a Technological Regime That Develops Genetic
Engineering but Locks Out Agroecological Innovations," *Research
Policy* 38, no. 6 (2009): 971–83.

Varshney, Rajeev K., Jean-Marcel Ribaut, Edward S. Buckler, Roberto Tuberosa, J. Antoni Rafalski, and Peter Langridge. "Can Genomics Boost Productivity of Orphan Crops?" *Nature Biotechnology* 30, no. 12 (2012): 1172. http://doi.org/http://doi.org/10.1038/nbt.2440.

Vetten, H. Joseph, Ute Hoyer, Edgar Maiss, Dietrich E. Lesemann, and Wilhelm Jelkmann. "Serological Detection and Discrimination of Geographically Diverse Isolates of Sweet Potato Sunken Vein Closterovirus." *Phytopathology* 86, no. 11 (1996): 744–50.

Vieira, Martin. "Community-Led Approach Sustains Livelihood Improvement in Kotoba, Ethiopia." In *Agroecology: The Bold Future of Farming in Africa*, edited by Michael Farrelly, G. Clare Westwood, and Stephen Boustred, 28–31. Dar es Salaam, Tanzania: Alliance for Food Sovereignty in Africa and Tanzania Organic Agriculture Movement, 2016. http://afsafrica.org/wp-content/uploads/2017/02/Agroecology-the-bold-future-of-farming-in-Africa-ebook-2-page-spreads.pdf.

Viljoen, Chris, and Lukeshni Chetty. "A Case Study of GM Maize Gene Flow in South Africa." *Environmental Science Europe* 23 (2011): 8–15. http://doi.org/10.1186/2190-4715-23-8.

Vitale, Jeffrey, and John Greenplate. "The Role of Biotechnology in Sustainable Agriculture of the Twenty-First Century: The Commercial Introduction of Bollgard II in Burkina Faso." In *Convergence of Food Security, Energy Security and Sustainable Agriculture*, edited by David D. Songstad, Jerry L. Hatfield, Dwight T. Tomes, 239–93. Heidelberg: Springer, 2014.

Vitale, Jeffrey, Marc Ouattarra, and Gaspard Vognan. "Enhancing Sustainability of Cotton Production Systems in West Africa: A Summary of Empirical Evidence from Burkina Faso." *Sustainability* 3, no. 8 (2011): 1,136–69. http://doi.org/10.3390/su3081136.

Vitale, Jeffrey, Gaspard Vognan, Marc Ouattarra, and Ouola Traoré. "The Commercial Application of GMO Crops in Africa: Burkina Faso's Decade of Experience with Bt Cotton." *AgBioForum*, 13, no. 4 (2010): 320–32. ISSN: 1522-936X.

Vitale, Jeffrey, Harvey Glick, John Greenplate, Mourad Abdennadher, and Oula Traoré. "Second-Generation Bt Cotton Field Trials in Burkina Faso: Analyzing the Potential Benefits to West African Farmers." *Crop Science* 48 (2008): 1958–66. http://doi.org/0.2135/cropsci2008.01.0024.

Voytas, Daniel F., Caixia Gao, and S. McCouch. "Precision Genome Engineering and Agriculture: Opportunities and Regulatory Challenges."

PLoS Biology 12, no. 6 (2014): e1001877. http://doi.org/10.1371/
journal.pbio.1001877.

Waddington, Stephen, M. Mudhara, M. Hlatshwayo, and P. Kunjeku.
"Extent and Causes of Low Yield in Maize Planted Late by Smallholder
Farmers in Sub-Humid Areas of Zimbabwe." *Farming Systems Bulletin,
Eastern and Southern Africa* 9 (1991): 15–31.

Waithaka, Michael M., Philip K. Thornton, Keith D. Shepherd, and
Nicholas N. Ndiwa. "Factors Affecting the Use of Fertilizers and
Manure By Smallholders: The Case of Vihiga, Western Kenya." *Nutrient
Cycling in Agroecosystems* 78 (2007): 211–24. http://doi.org/10.1007/
s10705-006-9087-x.

Waltz, Emily. "CRISPR-Edited Crops Free to Enter Market, Skip
Regulation." *Nature Biotechnology* 34, no. 6 (2016): 582. http://doi.
org/10.1038/nbt0616-582.

Wamboga-Mugirya, Peter. "Biotech Can Improve Agric Productivity."
BioVision Newsletter, September, no. 15 (2011): 5. http://www.
biovisioneastafrica.com/publications/biovision-15.pdf.

Wambugu, Florence. "Why Africa Needs Agricultural Biotech." *Nature*
400, no. 6,739 (1999): 15–16. http://doi.org/10.1038/21771.

– "Development and Transfer of Genetically Modified Virus-Resistant
Sweet Potato for Subsistence Farmers in Kenya." *Nutrition Reviews* 61,
no. 6, pt. 2 (2003): S110–13. ISSN: 0029-6643.

– "The Importance of Political Will in Contributions of Agricultural
Biotechnology towards Economic Growth, Food and Nutritional
Security in Africa." In *Biotechnology in Africa: Emergence, Initiatives
and Future*, edited by Florence Wambugu and Daniel Kamanga, 1–11.
Basel, Switzerland: Springer, 2014.

– "The Importance of Political Will in Contributions of Agricultural
Biotechnology towards Economic Growth, Food and Nutritional
Security in Africa." *Biotechnology in Africa: Emergence, Initiatives and
Future*, edited by Florence Wambugu and Daniel Kamanga. Basel,
Switzerland: Springer, 2014.

Wamuswa, Nanjinia. "New Maize Variety Gives Farmers Hope Against
Deadly Armyworm." *Standard Digital*, 14 March 2017. https://www.
standardmedia.co.ke/article/2001232639/new-maize-variety-gives-
farmers-hope-against-deadly-armyworm.

Waturu, Charles N. "The Status of Bt-Cotton Confined Field Trials in
Kenya." Powerpoint presentation, n.d. http://www.powershow.com/
view/1c78ad-OWYoM/Dr_Charles_N_Waturu_powerpoint_ppt_
presentation.

Wedding, Kristin, and Johanna Nesseth Tuttle. *Pathways to Productivity: The Role of GMOs for Food Security in Kenya, Tanzania and Uganda.* Washington, DC: Centre for Strategic and International Studies, 2013. http://csis.org/files/publication/131010_Wedding_PathwaysTo Productivity_Web.pdf.

Wezel, Alexander, Stephane Bellon, T. Doré, Charles Francis, Dominique Vallod, and David Christophe. "Agroecology as a Science, a Movement or a Practice. A Review." *Agronomy for Sustainable Development* 29 (2009): 503–15. http://doi.org/10.1051/agro/2009004.

Whitfield, Stephen. *Adapting to Climate Uncertainty in African Agriculture: Narratives and Knowledge Politics.* Abingdon, UK: Routledge, 2016.

Whitfield, Stephen, Jami Dixon, Brian Mulenga, and Hambulo Ngoma. "Conceptualising Farming Systems for Agricultural Development Research: Cases from Eastern and Southern Africa." *Agricultural Systems* 133 (2015): 54–61. http://doi.org/10.1016/j.agsy.2014.09.005.

Witcombe, John R., Krishna D. Joshi, Sanjaya Gyawali, A.M. Musa, C. Johansen, Daljit Virk, and Bhuwon Sthapit. "Participatory Plant Breeding Is Better Described as Highly Client-Oriented Plant Breeding: I. Four Indicators of Client-Orientation in Plant Breeding." *Experimental Agriculture* 41, no. 3 (2005): 299–319. http://doi.org/ 10.1017/S0014479705002656.

Witt, Harald, Rajeev Patel, and Matthew A. Schnurr. "Can the Poor Help Crops? Technology, Representation and Cotton in the Makhathini Flats, South Africa." *Review of African Political Economy* 33, no. 109 (2006): 497–513. http://www.jstor.org/stable/4007055.

World Bank. *World Development Report 2008: Agriculture for Development.* Washington, DC: World Bank, 2007.

– *Uganda: Promoting Inclusive Growth. Transforming Farms, Human Capital, and Economic Geography.* Washington, DC: International Bank for Reconstruction and Development, 2012.

– *Unlocking Africa's Agricultural Potential: An Action Agenda for Transformation.* World Bank Sustainable Development Series. Washington, DC: World Bank, 2013.

World Economic Forum and A.T. Kearney. *Grow Africa: Partnering to Achieve African Agricultural Transformation.* Geneva: World Economic Forum, 2016.

Wu, Ray, Wei Lin Guo, Xue Ren Wang, Xiu Lin Wang, Ting Tin Zhuang, Jihong Liu Clarke, and Bohan Liu. "Unintended Consequence of Plant Transformation: Biolistic Transformation Caused Transpositional

Activation of an Endogenous Retrotransposon *Tos17* in Rice ssp. *japonica* cv. Matsumae." *Plant Cell Reports* 28, no. 7 (2009): 1043–51. http://doi.org/10.1007/s00299-009-0704-4.

Xiong, Jin-Song, Jing Ding, and Yi Li. "Genome-Editing Technologies and their Potential Application in Horticultural Crop Breeding." *Horticulture Research* 2 (2015): 15019. http://doi.org/10.1038/hortres.2015.19.

Xu, Yunbi, and Jonathan Crouch. "Marker-Assisted Selection in Plant Breeding: From Publications to Practice." *Crop Science* 48, no. 2 (2008): 391–407. http://doi.org/10.2135/cropsci2007.04.0191.

Yapa, Lakshman. "Improved Seeds and Constructed Scarcity." In *Liberation Ecologies: Environment, Development and Social Movement*, edited by Richard Peet and Micheal Watts. New York: Routledge, 2002.

Yates, Mary Carlin. "Ghana: Request for Funds for Biotechnology Outreach Programs." Wikileaks Cables: 05ACCRA110_a. Dated 14 January 2005. https://wikileaks.org/plusd/cables/05ACCRA110_a.html.

Yirgou D., and J.F. Bradbury. "A Note on Wilt of Banana Caused by the Enset Wilt Organism Xanthomonas Musacearum." *Agricultural and Forestry Journal* 40 (1974): 111–14. http://doi.org/10.1080/00128325.1974.11662720.

Zaid, Abdelouahhab, Hughes G. Harrison, Enrico Porceddu, and Frank Nicholas. *Glossary of Biotechnology for Food and Agriculture: A Revised and Augmented Edition of the Glossary of Biotechnology and Genetic Engineering*. Rome: Food and Agriculture Organization of the United Nations (FAO), 2001.

Zerbe, Noah. "Seeds of Hope, Seeds of Despair: Towards a Political Economy of the Seed Industry in Southern Africa." *Third World Quarterly* 22, no. 4 (2001): 657–73. http://doi.org/10.1080/01436590120071830.

– "Feeding the Famine? American Food Aid and the GMO Debate in Southern Africa." *Food Policy* 29, no. 6 (2004): 593–608. http://doi.org/10.1016/j.foodpol.2004.09.002.

– "Biodiversity, Ownership, and Indigenous Knowledge: Exploring Legal Frameworks for Community, Farmers, and Intellectual Property Rights in Africa." *Ecological Economics* 53, no. 4 (2005): 493–506. http://doi.org/10.1016/j.ecolecon.2004.10.015.

– "Contesting Privatization: NGOs and Farmers' Rights in the African Model Law." *Global Environmental Politics* 7, no. 1 (2007): 97–119. http://doi.org/10.1162/glep.2007.7.1.97.

Zhu, April. "Calestous Juma on Being Pro-Africa, Why Africa Needs GM Crops, and How He Came to Be a Cheerleader." *Huffington Post*, 2 July 2014. https://www.huffingtonpost.com/april-zhu/let-us-innovate-calestous_b_5249823.html.

Zimmerer, Karl S., and Thomas J. Bassett. "Approaching Political Ecology: Society, Nature, and Scale in Human-Environmental Studies." In *Political Ecology: An Integrative Approach to Geography and Environment-Development Studies*, edited by Karl S. Zimmerer and Thomas J. Bassett, 1–25. New York: Guilford Press, 2003.

Index

stem borers, 127, 141, 145, 159,
163–4; and alternative management strategies, 121; and cultural
methods of control, 169; damages
by, 114–15, 121, 124, 144, 158–
9, 232n13; local farmer knowledge of, 119; precipitation effects
on, 118–19; and refugia, 122
stress tolerance, 32, 45, 126
Stress Tolerant Maize for Africa
(STMA), 156
sub-Saharan Africa (SSA), 30, 32–3,
35–7, 152–5, 158–9; absence of
dormancy period in, 114; agricultural productivity in, 3, 28;
average land holdings in, 240n7;
farm density in, 242n54
subsistence farming, 6–8, 10, 115,
117; and family labour, 191; and
intercropping practices, 71, 99,
103; and nutrition, 180; and
race, 113
Sudan, 187, 225n101, 228n75; as
biosafety regulation renegade,
41–2, 62–3; and cotton, 22, 69,
91–3, 97, 107, 196
surveys, 116, 147, 158, 177; demographic, 180; health, 180; large-
scale, 107, 198–9, 201; pest, 103,
121; refugia, 123; transect, 161
sustainable intensification, 4, 12,
207
Swaziland, 42, 63, 204
sweet potato, 6, 22, 33, 140–1. See
also GM sweet potato
sweet potato feathery mottle virus
(SPFMV), 139–40, 145–6, 148
Syngenta Foundation for Sustainable Agriculture (SFSA), 127,
141–5, 149

synthetic biology, 38
systems-agronomy approach,
207–8

Tanzania, 156, 159–62, 170, 182,
187, 241n20; as biosafety regulatory resister, 41–2, 56–60, 204
technological narratives, 21, 72,
202; and myth of success, 70,
77–8, 84–5, 107, 113–28; as
overpowering smallholder stories, 204
technology co-development, 44–5,
207, 213, 220n16
technology fees, 115, 119, 125,
127; as adoption challenge,
75–6, 102, 105, 231n137,
234n38; and smallholder economic risk, 118, 121
technology pipeline, 156, 169, 201,
206
technology transfer, 44, 62, 138,
207, 211, 213
tissue culture, 33–4, 140, 188–90,
192–3, 199, 212
tobacco, 7, 8, 169
transgenes, 34, 125, 148
transgenesis, 147, 149–50, 157–8,
180, 202, 250n83; as compared
to conventional plant breeding,
34–5; and manipulation of target
genes, 34, 37; as escalating resistance strategy, 125
transgenic resistance, 132
transgenic treadmill, 126

Uganda, 6, 16, 203–4, 209,
241n20, 245n6, 248n60; as biosafety regulation emerging
adopter, 41–2, 48–51, 54, 57–9,

109–13, 240n6, 240n7; and
matooke, 173, 173–6, 178–81,
185–91; potential, 3, 7, 195,
241n24; and potential for buyer
contracts, 190; stable, 111, 120,
133, 190; sweet potato, 140–1,
146–8, 150; and WEMA, 153–9,
165–7

Zambia, 153, 171, 223n67,
232n11, 237n97, 242–3n54; as
biosafety regulatory resister,
41–2, 54–5, 59–60; and food aid
importation, 56–7, 204; GM
detection lab in, 56
Zimbabwe, 41–2, 56–7, 59–60,
111, 204, 232n11